This book contains materials developed by the AIMS Education Foundation. **AIMS** (**A**ctivities **I**ntegrating **M**athematics and **S**cience) began in 1981 with a grant from the National Science Foundation. The non-profit AIMS Education Foundation publishes hands-on instructional materials (books and the monthly magazine) that integrate curricular disciplines such as mathematics, science, language arts, and social studies. The Foundation sponsors a national program of professional development through which educators may gain both an understanding of the AIMS philosophy and expertise in teaching by integrated, hands-on methods.

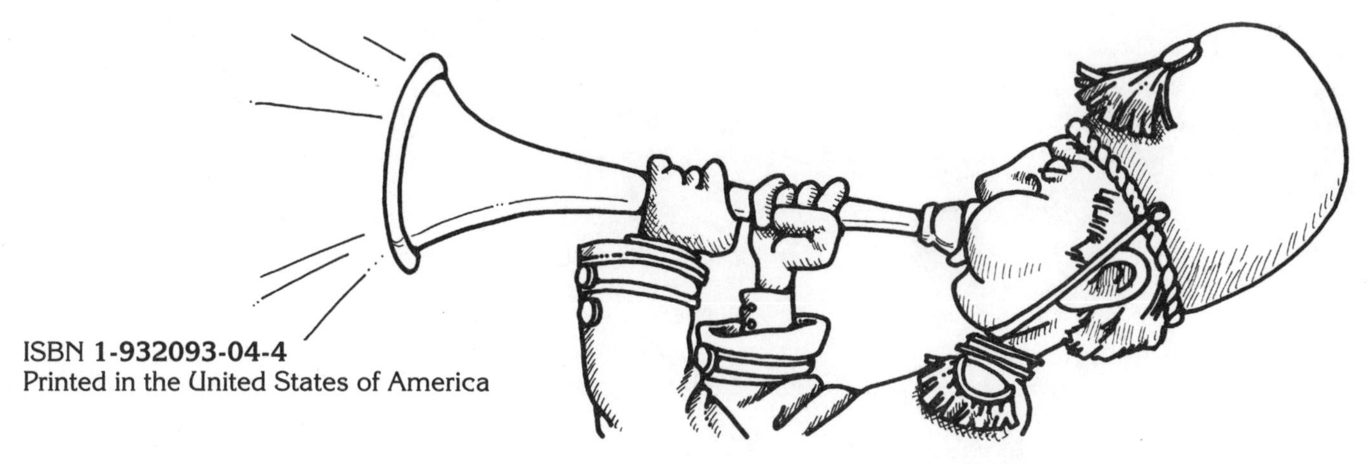

ISBN **1-932093-04-4**
Printed in the United States of America

LOOKING at GEOMETRY

Principal Author
Sheldon Erickson

Contributing Authors
Richard Thiessen
Ann Wiebe

Editor
Betty Cordel

Illustrator
Reneé Mason

Desktop Publisher
Tracey Lieder

I Hear and
I Forget,
I See and I
Remember,
I Do and I
Understand.
Chinese Proverb

LOOKING at GEOMETRY

Looking at Formulas

Looking at Dimensions

Patterns, Problem Solving, and Practice

NCTM Standards 2000*

Geometry

Analyze characteristics and properties of two- and three-dimensional geometric shapes and develop mathematical arguments about geometric relationships

- Precisely describe, classify, and understand relationships among types of two- and three-dimensional objects (e.g., angles, triangles, quadrilaterals, cylinders, cones) using their defining properties
- Understand relationships among the angles, side lengths, perimeters, areas, and volumes of similar objects
- Create and critique inductive and deductive arguments concerning geometric ideas and relationships, such as congruence, similarity, and the Pythagorean relationship

Apply transformations and use symmetry to analyze mathematical situations

- Describe sizes, positions, and orientations of shapes under informal transformations such as flips, turns, slides, and scaling

Use visualization, spatial reasoning, and geometric modeling to solve problems

- Draw geometric objects with specified properties, such as side lengths or angle measures
- Use two-dimensional representations of three-dimensional objects to visualize and solve problems such as those involving surface area and volume
- Use geometric models to represent and explain numerical and algebraic relationships
- Recognize and apply geometric ideas and relationships in areas outside the mathematics classroom, such as art, science, and everyday life

Measurement

Understand measurable attributes of objects and the units, systems, and processes of measurement

- Understand, select, and use units of appropriate size and type to measure angles, perimeter, area, surface area, and volume

Apply appropriate techniques, tools, and formulas to determine measurements

- Select and apply techniques and tools to accurately find length, area, volume, and angle measures to appropriate levels of precision
- Develop and use formulas to determine the circumference of circles and the area of triangles, parallelograms, trapezoids, and circles and develop strategies to find the area of more-complex shapes
- Develop strategies to determine the surface area and volume of selected prisms, pyramids, and cylinders

Algebra

Understand patterns, relations, and functions

- Represent, analyze, and generalize a variety of patterns with tables, graphs, words, and, when possible, symbolic rules
- Relate and compare different forms of representation for a relationship
- Identify functions as linear or nonlinear and contrast their properties from tables, graphs, or equation

Represent and analyze mathematical situations and structures using algebraic symbols

- Develop an initial conceptual understanding of different uses of variables
- Use symbolic algebra to represent situations and to solve problems, especially those that involve linear relationships

Use mathematical models to represent and understand quantitative relationships

- Model and solve contextualized problems using various representations, such as graphs, tables, and equations

Analyze change in various concepts

- Use graphs to analyze the nature of changes in quantities in linear relationships

LOOKING AT FORMULAS

Developing and Using Formulas

Geometry and measurement are intrinsically related. At the middle-school level, measurement focuses around the different dimensions of length, area, and volume. Difficulty arises when area and volume are determined indirectly using formulas derived from linear measures. For example, rather than understanding the concept of area as a covering, students simply rely on the formula of length times width. Researchers report that when students are given two dimensions of a rectangle and are asked for the perimeter, they often multiply the dimensions using the area formula. If the dimensions of four sides of the rectangle are labeled, students often calculate the perimeter when the question concerns area. They are applying the formulas incorrectly because they have not gained an understanding of the concepts of measurement.

Students need multiple experiences using a variety of manipulatives to gain an understanding of the concepts of measurement and to construct the formulas for themselves. The experiences in this section will look at how teachers might use Geoboards, Algebra Tiles, paper cutting, blocks, investigations, and puzzles to develop the concepts and the formulas.

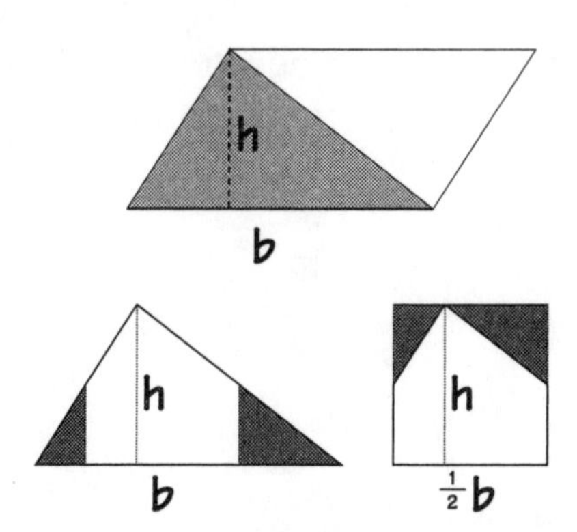

The variety of experiences provides for different learning styles and reinforces learning from the past experiences. Each experience encourages the development of a formula. Multiple experiences produce a variety of formulas for one concept that, although equivalent, provide different perspectives.

Consider the following examples of multiple experiences: As students study the area of triangles with Geoboards, they generally surround the triangle they are studying with a parallelogram and find half of its area. The resulting equation of one-half the quantity of the base times the height [$A = \frac{1}{2}(b \bullet h)$] reflects the experience. The product of the base and the height is the area of the parallelogram, and half of it is the triangle.

The experience of cutting a paper triangle provides a different perspective on the area formula. Students cut the triangle at its altitude and then cut the base of the resulting right triangles in half forming small right triangles and trapezoids. The two small right triangles are rotated to form rectangles with the trapezoids. The two rectangles form a large rectangle that is the same height as the original triangle but is only half as wide. The descriptive formula for this experience is the product of half the base and the height [$A = \frac{1}{2}b \bullet h$].

The perspectives provided by numerous and varied experiences develop a rich and broad understanding. When students use the formulas they have constructed to find missing components, their understanding of geometry is reinforced. By working backwards, they also develop algebraic thinking. Consider the situation that a triangle has an area of 25 square units and is five units high. Experienced students will visualize the triangle and double it to get the parallelogram that surrounded it on the Geoboard. Since they now know the area of the parallelogram is 50 square units and one of its factors is five, they divide to get the missing factor of ten, the length of the base. Although they have not come to their solutions using symbols, they have done the algebraic thinking required to get the solution. It is a simple step for the teacher to help the students symbolize what they have done:

understand the situation as a formula	=	$A = \frac{1}{2}(b \bullet h)$
substitute known quantities	=	$25 = \frac{1}{2}(b \bullet 5)$
double area to get parallelogram	=	$25 \bullet 2 = \frac{1}{2}(b \bullet 5) \bullet 2$
		$50 = b \bullet 5$
divide by 5 to get the missing factor	=	$50 \div 5 = b \bullet 5 \div 5$
		$10 = b$

The middle-school student requires numerous experiences in a variety of contexts to develop and understand geometric formulas. Time spent in these experiences develops intuitive understanding of concepts of measurement and the relationship of formulae to measurements. The skilled teacher will utilize these concrete experiences to help students construct meaningful bridges to their symbolic representations as formulas developing greater geometric and algebraic understandings.

Background Article

The Pythagorean Relationship Part One

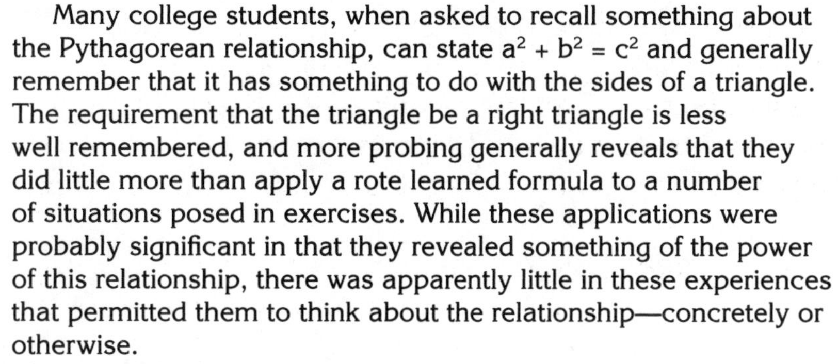

Many college students, when asked to recall something about the Pythagorean relationship, can state $a^2 + b^2 = c^2$ and generally remember that it has something to do with the sides of a triangle. The requirement that the triangle be a right triangle is less well remembered, and more probing generally reveals that they did little more than apply a rote learned formula to a number of situations posed in exercises. While these applications were probably significant in that they revealed something of the power of this relationship, there was apparently little in these experiences that permitted them to think about the relationship—concretely or otherwise.

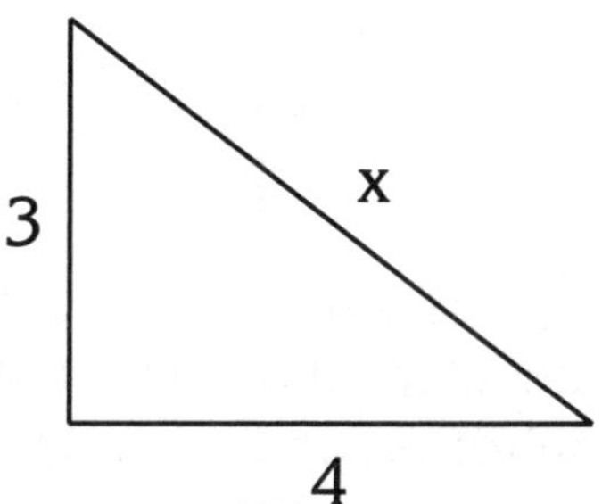

That this would be the case is not surprising when you consider a typical textbook presentation of the Pythagorean relationship. A current, widely used textbook for junior high introduces the relationship with the following sentence, "When you know the lengths of two sides of a right triangle, you can use the Rule of Pythagoras to find the length of the other side." This is followed by a statement of the formula and a three step procedure for applying the rule. After two examples, students are asked to do several exercises that picture triangles in which the lengths of two sides are given and students are asked to find the length of the third side.

This article, as well as the next, will suggest some activities and areas of exploration that might be used with students at various levels to help them build adequate mental structures for this relationship. These experiences can serve as a basis for thought about the relationship; to make it meaningful and plausible for them.

While the relationship is attributed to Pythagoras, specific instances of it were known much earlier by both the Egyptians and Babylonians. A story about ancient Egypt relates how "rope stretchers" used ropes with equally spaced knots to re-establish the boundaries that separated one man's property from his neighbor's after the waters of the Nile River overflowed their banks each spring. As the story goes, these early surveyors established right triangles with side measures of three, four, and five units using the knots in a rope. This permitted them to construct the right angles needed to determine property boundaries of rectangular plots. Although they knew about and used this one instance of the relationship, it probably occurred to them to look for or think about other number triples that would also yield a right triangle.

This story and the method used can provide an opportunity for students to begin to explore and experience the Pythagorean relationship. Using some string of sufficient length (12 feet or so works well), tie loops at approximately 12-inch intervals. If 13 such loops are tied and excess string is snipped off at either end, you have 13 loops with 12 equal spaces. Now tie the first and last loops together and consider them as one. This gives you 12 equally spaced loops and 12 equal spaces.

The loops are very convenient as places to hold onto the string, or a pencil can be slipped through the loop as a way to hold onto the string.

Now have a student hold onto the string at one of the loops, count three spaces and have another student hold that loop; finally, count four more spaces and have a third student hold that loop. You now have three, four, and five spaces, respectively, between the loops being held by students. If the students now stretch the string between each of them, the resulting triangle has sides of lengths three, four, and five with a right angle at the vertex opposite the side having length of five. A corner of a sheet of paper or a book can be used to check the right angle.

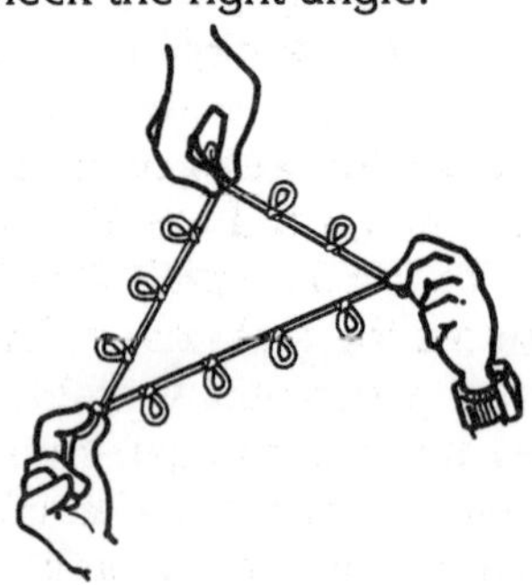

Students can be asked to consider why three, four, and five were chosen as the numbers of units for the sides of the triangle. They can further be asked to try some other possibilities. What happens if each of the numbers is multiplied by two or three? Will the resulting triangles of sides six, eight, 10, and nine, 12, 15 also yield right triangles? Or how about trying five, 12, and 13 as sides? For this exploration it would be helpful to have additional strings with appropriate numbers of loops. Students should also try some other lengths of sides of their own choosing to see that this does not happen for just any three lengths. For example, suppose the sides are all the same. This could be checked using the cord with 12 units by simply placing the pencils so that there are four spaces between pencils. The 12-unit cord could also be used to check a number of other possibilities as well. Activities such as this allow students to begin thinking inductively about the Pythagorean relationship as opposed to having it stated as simply a rule that is to be applied.

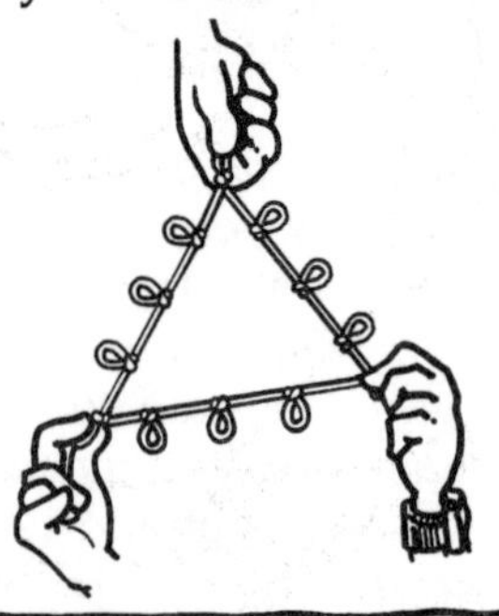

A related activity to permit additional exploration of the relationship can be devised using geostrips of various lengths. Suppose students are provided with strips of lengths two, three, four, and five, and so on up to 20. Students can then be asked to try combinations of sides of triangles that appear to yield a right triangle. Again, the activity provides for exploration and permits children to again notice that not all combinations of sides produce a right triangle. Furthermore, it gives opportunity for them to wonder about the conditions that yield right triangles.

It is unlikely that students will reach the conclusion that the relationship, $a^2 + b^2 = c^2$, is always satisfied when the triangle formed is a right triangle. Giving them the relationship after having experiences like the ones described, and having them apply it to a number of instances will certainly help to give meaning to the relationship. They will probably find that in some cases triangles they thought appeared to be right were in fact only close to having a right angle. Applying the formula provides them a way to check their perceptions. For example, a triangle with sides of 10, 12, and 15 may appear to be a right triangle; however, the relationship is not satisfied.

Having concluded that the formula is only satisfied for right triangles, we might ask about the nature of the triangles where $a^2 + b^2$ is not equal to c^2. Clearly there are two possibilities for the relationship between lengths of sides, either $a^2 + b^2 < c^2$ or $a^2 + b^2 > c^2$. Furthermore, there are two possibilities as to classification of the triangles; either they are acute or they are obtuse. Let's go back to the 30-loop string that we used to test the 5, 12, 13 triangle. Consider the triangle with sides 14, six, and 10. What kind of triangle? It can't be a right triangle, so it is either obtuse or acute. Which is it? Remember that we always take c as the longest side. What happens with $a^2 + b^2$ and c^2? Which is greatest? By exploring a variety of possibilities, students should conclude that if a triangle is acute, then $a^2 + b^2 > c^2$, and if it is obtuse, then $a^2 + b^2 < c^2$.

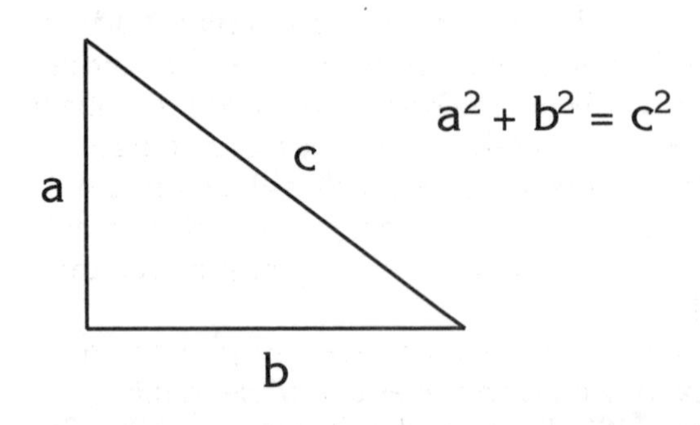

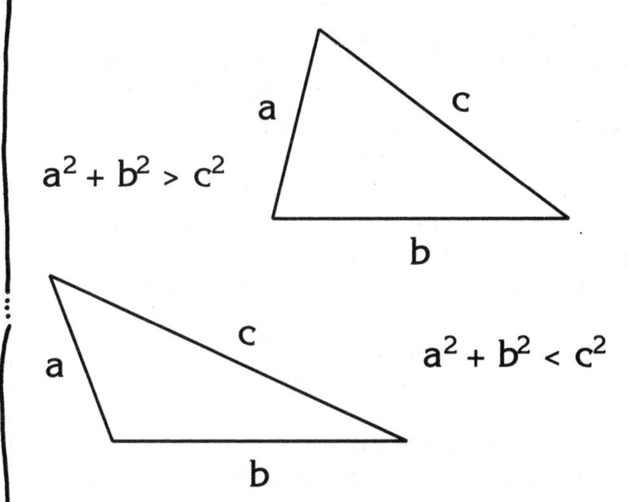

Activities such as these are important in helping students understand that the formula only applies when the triangle is a right triangle, and conversely, that when the formula is satisfied, then the triangle must be a right triangle. This is a connection that is too often missed or at least not well understood by students.

We have restricted our activities to situations involving only whole numbers. This seems reasonable in that we do not assume that students at this level have experienced irrational numbers and so are not ready to deal with triples like two, three, 13. There is a nice extension of what we have done so far that would be appropriate for students at this level and would permit further exploration of the relationship as well as provide further opportunity to involve students in inductive reasoning activities.

Triples of whole numbers that satisfy $a^2 + b^2 = c^2$ are often called Pythagorean triples. Consider the following list of triples as an example. A couple of these triples should look familiar; however, most will be new. Can you find the missing numbers?

3, 4, and 5	11, 60, and 61
5, 12, and ___	___, ___, and 85
7, 24, and ___	___, ___, and ___
___, 40, and 41	

Looking at the pattern it is clear that the first numbers of the triples are consecutive odd numbers. Furthermore, the pattern reveals that the other two numbers are consecutive whole numbers with the first being even. Noticing these patterns makes it possible to fill all the blanks except the last two. Do you see any other patterns that might be helpful? You may want the help of a calculator this time.

Now that you have completed the pattern so far, suppose the first number of a triple in this sequence is 97. What are the other two numbers? Would it help to know that they are greater than 4000?

If you have not discovered the pattern yet, let me give you a hint. Compare the square of the first number and double the second number. We will come back to this pattern in the next part.

The pattern expressed by the above sequence of Pythagorean triples can also be expressed as a set of simple formulas. We'll address that later, but for now, let's look at another set of formulas that also generate Pythagorean triples. Consider the following formulas where p is larger than q:

$$a = p^2 - q^2$$
$$b = 2pq$$
$$c = p^2 + q^2$$

Try some values for p and q. For example, what are a, b, and c, if p is 3 and q is 2? Are the numbers generated a Pythagorean triple? Which values of p and q will generate the triple three, four, and five? Which values will generate seven, 24, and 25? Can you find values of p and q to generate each of the triples in the earlier sequence? Can you generate some triples that are not on the previous list?

Since each of these triples satisfies the Pythagorean relationship, the three must represent lengths of sides of right triangles. Also, since there has been no reference to lengths that are not whole numbers, the relationship can be explored extensively without reference to numbers like the square root of five or other irrational numbers.

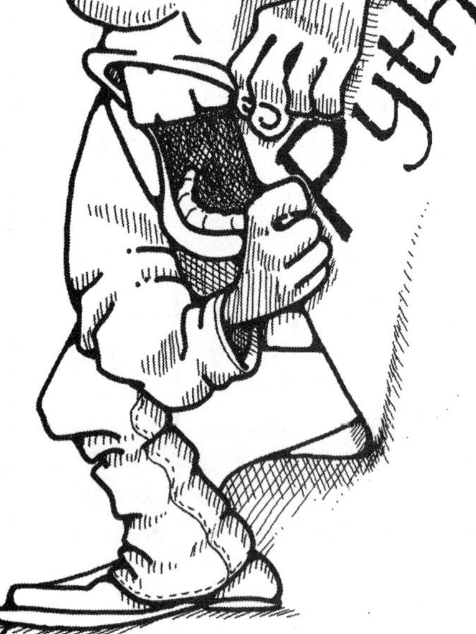

The Pythagorean Relationship

Part Two

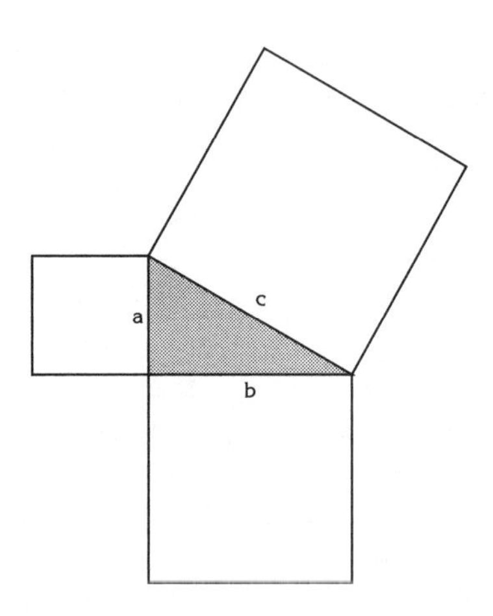

The Pythagorean relationship can be thought about concretely by considering a right triangle with squares constructed on each of its sides. The relationship then states that the sum of the areas of the two smaller squares is equal to the area of the larger square. This is, in fact, the way in which Pythagoras apparently thought about the relationship and it is the way in which he stated it.

Unfortunately, textbook presentations often omit any discussion of this aspect of the relationship, and as a result students miss out on a powerful way to picture and think concretely about it.

Using areas of squares constructed on the sides of triangles, let's revisit an activity we explored in *Part One*. In that activity, students explored various triangles and noticed that certain lengths of sides produce right triangles while other combinations produce triangles that are either acute or obtuse. We suggested that it would still be necessary for us to give them the formula, $a^2 + b^2 = c^2$, and have them notice that it holds for right triangles but not for others. Our suggestion was probably appropriate, given the student activities and our approach to that point. However, the following activity, which makes use of areas of squares constructed on the sides of a right triangle, opens the possibility for teachers to lead students to discover the formula on their own.

Give students centimeter graph paper and have them cut out a set of squares having areas of four, nine, 16, 25, 36, 49, 64, 81, 100, 121, 144, and 169. Ask students to form triangles using the sides of three of the triangles.

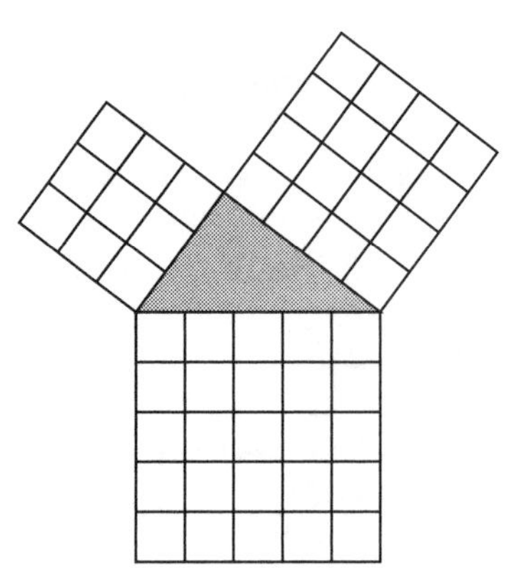

How many different right triangles can be constructed using various combinations of three of these squares? Is it also possible to construct triangles that are acute? Is it possible to construct triangles that are obtuse?

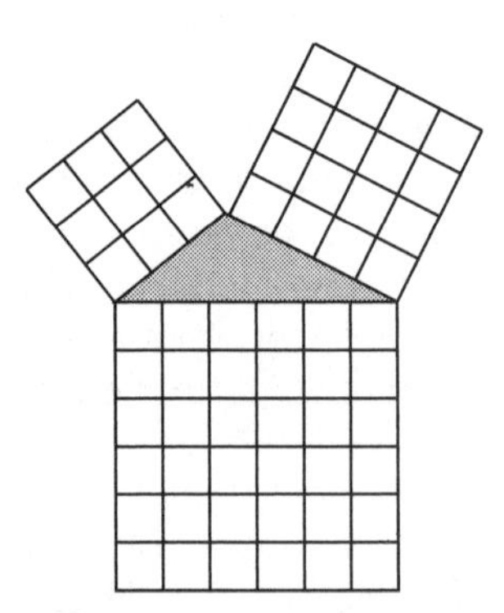

All of these are possible, of course, and as was done in the activity in *Part One*, students can then be asked about the relationship between the sum of the areas of the smaller two squares and the area of the largest square.

There are only three right triangles that can be constructed using the given squares; however, students can discover that for each of these triangles, the sum of the two smaller areas is equal to the larger area. Furthermore, they can notice by way of contrast that for all of the other triangles that can be constructed with these squares, and which are not right triangles, the sum of the areas of the two smaller squares is sometimes greater than and sometimes less than the area of the largest square.

Recognizing that the area of a square can be expressed in the form s^2, where s is the length of a side of a square, students can then be led to generalize the relationship expressed by the formula, $a^2 + b^2 = c^2$, for right triangles.

Students can further be led to generalize that for acute triangles the relationship can be expressed as $a^2 + b^2 > c^2$, and for obtuse triangles as $a^2 + b^2 < c^2$.

Thinking about the Pythagorean relationship in terms of areas of squares leads the way to a number of concrete, intuitive proofs of the relationship. The remainder of the article will be devoted to exploring some of these proofs. In each case we want to show that for any right triangle with squares constructed on each of its sides, the sum of the areas of the squares on the two shortest sides is equal to the area of the square on the longest side. (The longest side is called the hypotenuse and the two shorter sides are generally referred to as legs.)

As we work through the proofs, it will be helpful for you, the reader, to have triangles and squares to manipulate. A set of pieces that can be constructed from card stock or other heavy paper will be described for each of the proofs. However, since we will be using the same right triangle for each of the proofs, some of the pieces can be used for more than one proof. In fact, some of the same pieces can be used in each of the proofs.

For the first proof, (see *Left Behind—Pythagorean Proof 1*) construct eight congruent right triangles with sides of any length you want. Label the shorter leg of each triangle a; label the other leg b; and label the hypotenuse c. Now construct three squares having lengths of sides a, b, and c, respectively. Using these pieces, lay them out as pictured below.

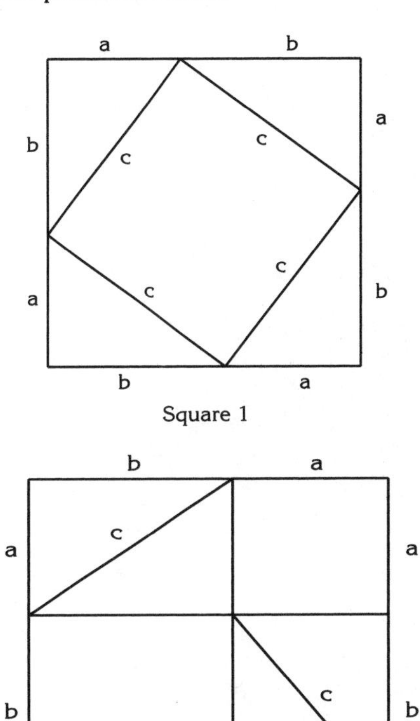

Square 1

Square 2

Two squares are formed each with a side of length a + b. Since the lengths of the sides of the two squares are the same, they must have the same area. In other words, the area of Square 1 is equal to the area of Square 2. Now suppose we removed the four triangles from each of the squares. Is what remains of Square 1 equal to what remains of Square 2? Since each of the triangles has the same area and we removed the same amount of area from each square, what remains of the two squares must be equal in area. But what is it that remains? In Square 1 it is a square with area c^2 and in Square 2 it is two squares having areas of a^2 and b^2, respectively. This, of course, is what we wanted to show.

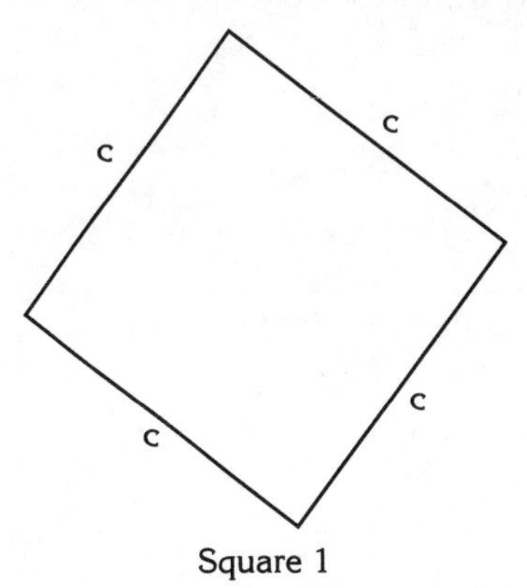

Square 1

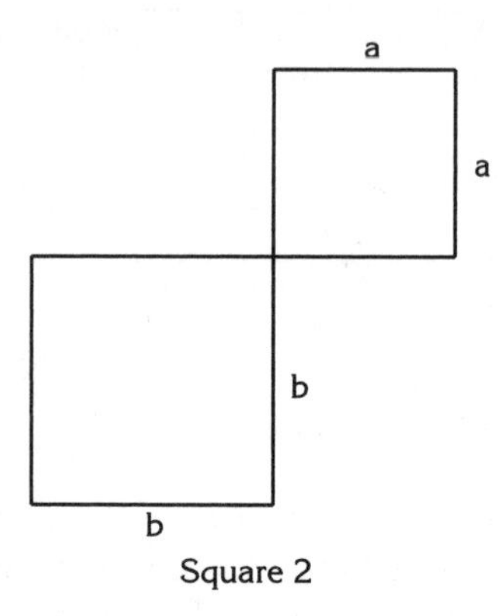

Square 2

While for some students this will be as far as we may want to go, others may be able to follow an algebraic proof of what we just did. Consider that for Square 1 we can express the area in two ways, and so we have the equation:

$$(a + b)^2 = 4(\frac{1}{2}ab) + c^2.$$

For Square 2 we can express the area in two ways as:

$$(a + b)^2 = 4(\frac{1}{2}ab) + a^2 + b^2.$$

One way to express the fact that Square 1 and Square 2 have equal area is by:

$$4(\frac{1}{2}ab) + a^2 + b^2 = 4(\frac{1}{2}ab) + c^2.$$

Subtracting $4(\frac{1}{2}ab)$, the total area of the four triangles, from both sides of the above equation we have:

$$a^2 + b^2 = c^2.$$

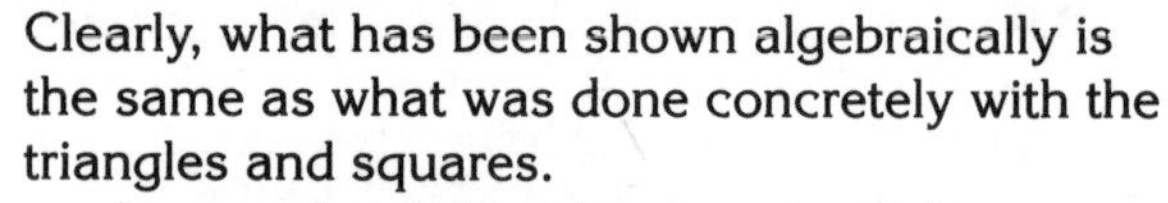

Clearly, what has been shown algebraically is the same as what was done concretely with the triangles and squares.

A second proof produces a puzzle (see *Pythagorean Puzzle*). The solution to the puzzle shows that the sum of the areas of the squares constructed on the legs of a right triangle is equal to the area of the square on the hypotenuse.

To construct the puzzle, start with a right triangle with squares constructed on only the two legs. Draw two lines through the square on the longer leg so that they intersect at right angles and one of the lines is perpendicular to the hypotenuse of the right triangle. Now cut out the two squares and also cut the larger square into four pieces along the perpendicular lines.

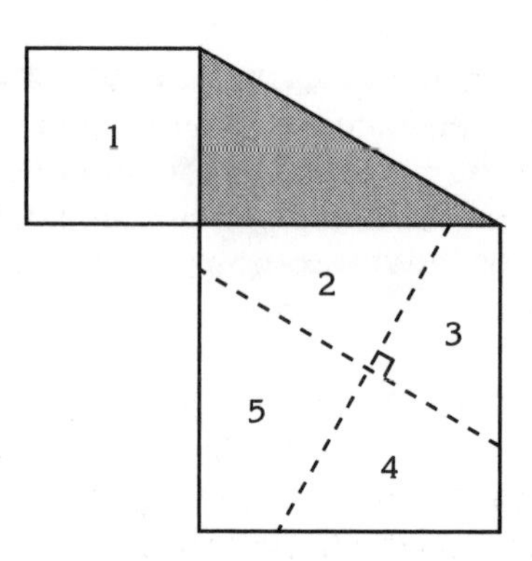

The puzzle now involves putting the four pieces of the larger square along with the smaller square together to form a large square on the hypotenuse. Two examples with the cuts made somewhat differently are pictured below.

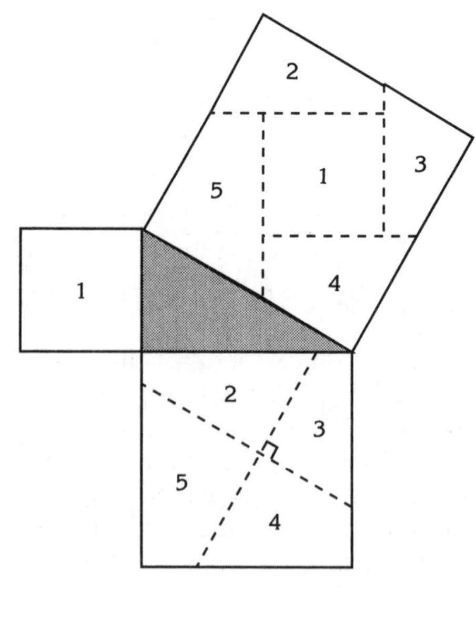

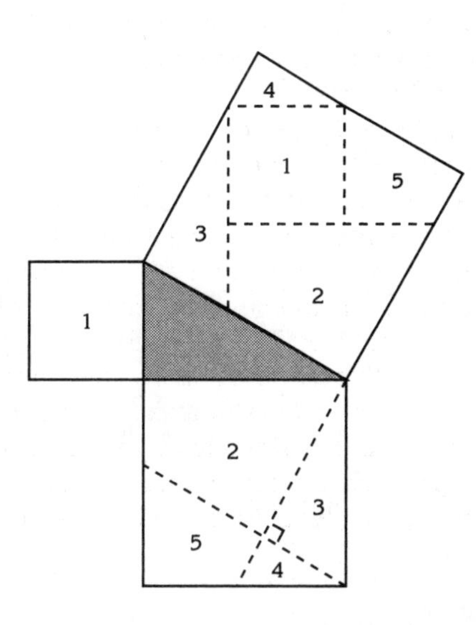

After constructing and solving the puzzle, students may be asked to explore the reasons for cutting the square on the longer leg as was done. Why make the first cut perpendicular to the hypotenuse? Why make the two cuts at right angles?

A simpler version of the puzzle can be made using an isosceles right triangle and cutting the squares on the equal sides along the diagonals. The resulting four pieces can be put together to form a square on the hypotenuse.

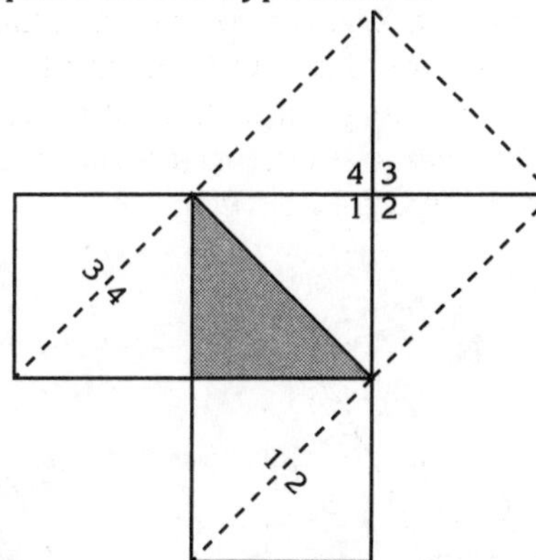

Two additional proofs are more algebraic; however, each can also be constructed using card stock pieces.

An ancient Hindu mathematician drew a diagram like the following and apparently gave no explanation other than to write above it the word "behold." The diagram simply shows four congruent right triangles arranged to form a square with each side being the hypotenuse of one of the triangles. (See *Squares and Triangles—Pythagorean Proof 2.*)

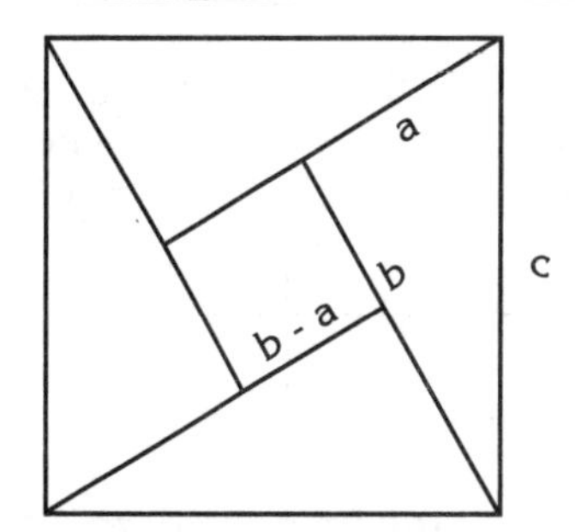

If you have cut out the triangles for the first proof, you can use them now to make your own arrangement of this square. The smaller square in the center has a length of side $b - a$.

The algebraic proof requires that we view the area of this square in two ways. Since the length of a side of the large square we constructed is c, the area of the square can be expressed as c^2. If we take into consideration the way in which the square was constructed, its area can be thought of as the sum of the areas of the four triangles and the area of the smaller square. This sum can be expressed as $4(\frac{1}{2}\ ab) + (b-a)^2$. Both expressions give the area of the square and so we have the following equation:

$$4(\frac{1}{2}\ ab) + (b - a)^2 = c^2$$

This equation simplifies to $a^2 + b^2 = c^2$, which is what we wanted to prove.

A final proof is attributed to President Garfield.(See *Garfield's Proof—Pythagorean Proof 3.*) He proved the relationship in 1876 while a member of the House of Representatives. The proof is based on the figure shown.

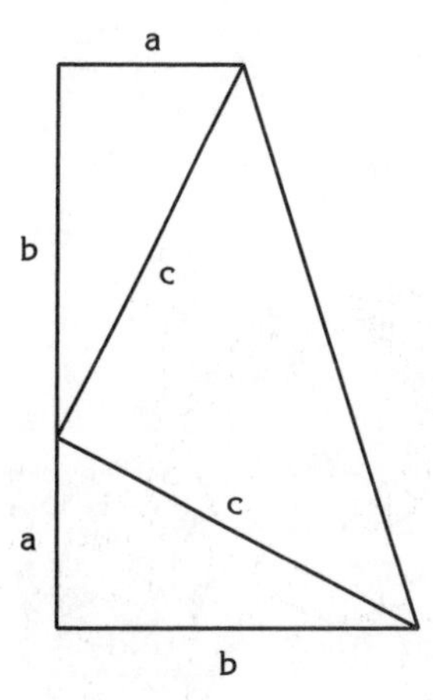

Three triangular pieces are arranged to form a trapezoid. The proof simply involves finding the area of the trapezoid in two ways as was done in the previous proof.

The formula for the area of a trapezoid is one-half the sum of the bases times the height. In this case the lengths of the bases are a and b, and the height is a + b. Using this formula, the area then can be expressed as:

$$\frac{1}{2}(a + b)(a + b).$$

If we find the area of the trapezoid as the sum of the areas of the three triangles that were used to make it, we have:

$$2(\frac{1}{2}ab) + \frac{1}{2}c^2.$$

Since the two expressions simply give the area of the trapezoid in two different ways, we can write the equation:

$$\frac{1}{2}(a + b)(a + b) = 2(\frac{1}{2}ab) + \frac{1}{2}c^2.$$

Simplifying this equation we have:

$$a^2 + b^2 = c^2$$

This is, of course, what we wished to prove.

While the last two proofs require a bit more algebra, using the pieces to construct the square and the trapezoid makes concrete the way in which the equations are formulated.

Part One started with the story of the rope stretchers who formed a right triangle using sides of three, four, and five. The explorations with the string of loops suggested other possible right triangles as well as triangles that were not right. These explorations would seem appropriate for students in middle school mathematics. They ended with proofs, some of which require the level of Algebra 1. Surely experiences like these are very different from those of the college student who can only recall having memorized the formula $a^2 + b^2 = c^2$

Finally, coming back to one of the sequences of Pythagorean triples: The first of each of the triples in the sequence was an odd number, while the last two numbers were consecutive whole numbers. The pattern was that to find the second number of the triple the first number needs to be squared, subtract one, and then divide by two. You can check to see that this works for the first seven triples. To find the answer for the triple with 97 as its first term, square 97 which is 9409. Now subtract one and divide by two. The answer is 4704 and the triple is 97, 4704, 4705.

A Pythagorean Puzzle

The Pythagorean theorem says that the sum of the areas of the squares of the two legs of a right triangle is equal to the area of the square of the hypotenuse.

Notice in the puzzle that there is a square on each leg of the right triangle. Cut out those two squares. Cut the bigger of the two squares into four pieces by cutting all the way across the square on any two lines that are perpendicular to each other.

Arrange the five pieces to make one square on the hypotenuse.

Explain how this illustrates the Pythagorean theorem.

1. Cut out the Pythagorean Proof Parts.

2. Cover squares 1 and 2 at the right with the parts listed.
   a. How long is the *edge* of each *square*?

   b. What is true about the areas of both squares?

   c. Write an *equation* by listing symbolically the parts in each square on either side of the *equal sign*.

3. Take away the four triangles from both squares.
   d. How do the *areas* of what is left in each *square* compare?

   e. Show the subtraction of the four triangles from both sides of your equation and find what remains.

4. Take the three squares and put them along the *corresponding edges* of one of the triangles.
   f. How does your final *equation* represent the *special relationship* that is true about the *edges of right triangles*?

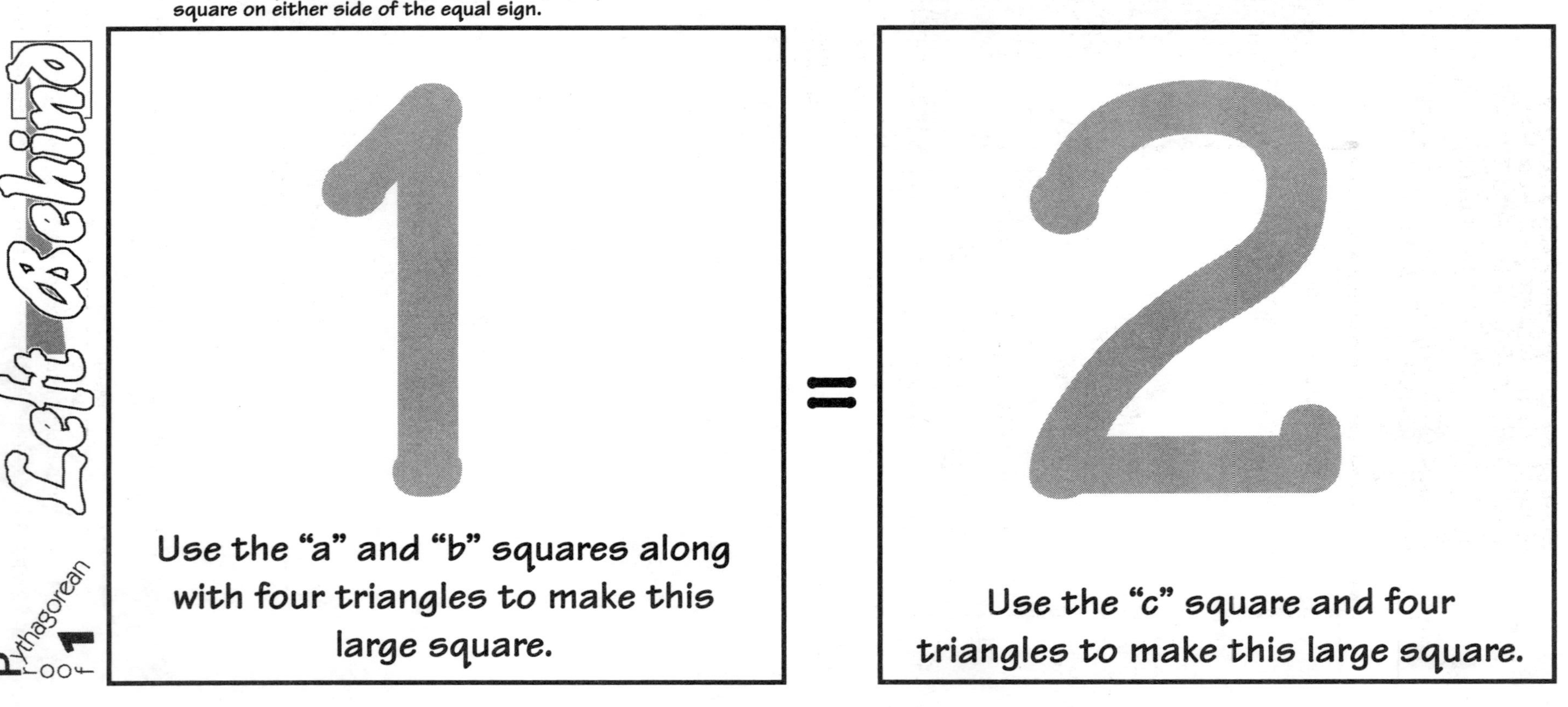

SQUARES AND TRIANGLES

Cut out the Pythagorean Proof Parts.

Use 4 triangles to fill this large square without covering the black square.

=

- What represents the length of the big square?
- Write the expression of how you would find the area of the big square.

The big square is made up of five parts.

- Write an expression of how you would find the area of one of the triangles.
- Write an expression that represents the length of the black square. Notice it is the difference between the "b" length and the "a" length.
- Write an expression of the area of the black square.
- The big square is made of four triangles and the black square. Fill in the areas of the parts to represent the area of the big square. Then simplify the expression.

4( ) + ( )

The two ways to represent the area of the large square are equal. Make an equation of the two expressions.

__________________ = __________________

Take the three squares (a, b, c) and put them along the corresponding edges of a single triangle. Explain how your equation represents the special relationship that is true about the edges of right triangles.

Pythagorean Proof 3

Garfield's Proof

Cut out the *Pythagorean Proof Parts.* Use two gray triangles and the one white triangle to fill this trapezoid.

- What represents the length of the top base(b_1)?
- What represents the length of the top base(b_2)?
- What represents the height (h)?
- Write the expression of how you would find the area of the trapezoid using the area formula. ($A = \frac{1}{2}(b_1 + b_2) \bullet h$) Simplify the expression.

The trapezoid is made up of three parts.

- Write an expression of how you would find the area of one of the gray triangles.
- Write an expression of how you would find the area of the white triangle.
- The trapezoid is made of two gray triangles and a white triangle. Fill in the areas of the parts to represent the area of the trapezoid. Then simplify the expression.

2(   ) + (   )

The two ways to represent the area of the trapezoid are equal. Make an equation of the two expressions.

________________ = ________________

Take the three squares (a, b, c) and put them along the corresponding edges of a single triangle. Explain how your equation represents the special relationship that is true about the edges of right triangles.

Pythagorean Proof Parts

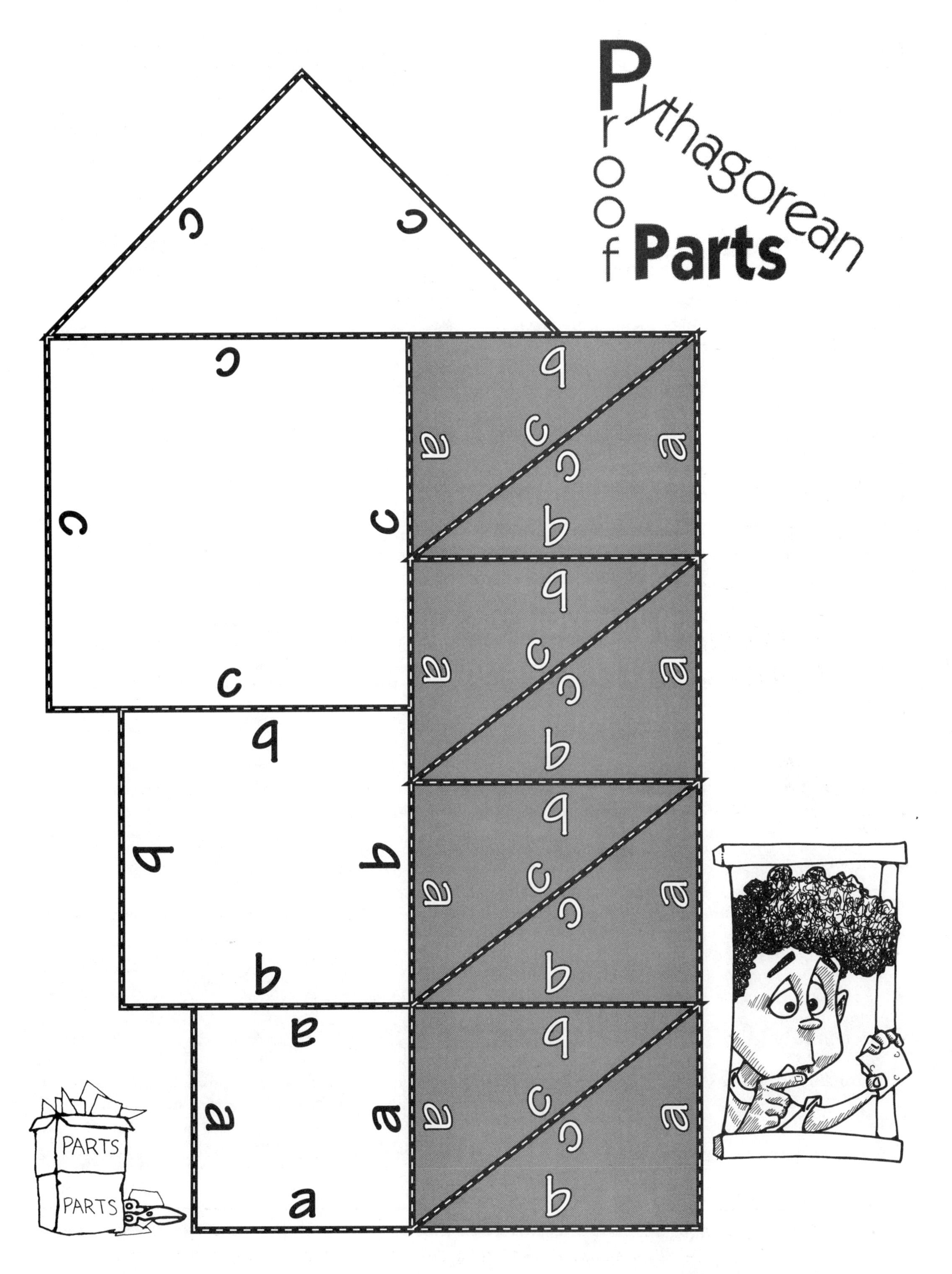

Geoboard Formulas

Topic
Measurement formulas

Key Question
How can you generalize a method of finding the areas of rectangles (parallelograms, triangles, trapezoids) by studying the figures (parallelograms, triangles, trapezoids) made on a Geoboard?

Learning Goals
Students will:
- develop an intuitive concept of perimeter and area of two-dimensional regions;
- measure perimeter and area of two-dimensional regions by counting;
- demonstrate visually ways to determine area by dissection and combining; and
- develop generalized formulas from their measuring experiences.

Guiding Documents
Project 2061 Benchmark
- *Calculate the circumferences and areas of rectangles, triangles, and circles, and the volumes of rectangular solids.*

*NCTM Standards 2000**
- *Understand relationships among the angles, side lengths, perimeters, areas, and volumes of similar objects*
- *Create and critique inductive and deductive arguments concerning geometric ideas and relationships, such as congruence, similarity, and the Pythagorean relationship*
- *Describe sizes, positions, and orientations of shapes under informal transformations such as flips, turns, slides, and scaling*
- *Use geometric models to represent and explain numerical and algebraic relationships*
- *Recognize and apply geometric ideas and relationships in areas outside the mathematics classroom, such as art, science, and everyday life*

Math
Geometry
  measurement (perimeter, area)
  Pythagorean theorem
Algebraic thinking
  developing formulae

Integrated Processes
Observing
Recording data
Generalizing

Materials
Geoboards
Rubber bands
Record sheets

General Overview of Materials
This set of materials has been split into five sections:
a. Rectangles
b. Parallelograms
c. Triangles
d. Trapezoids
e. Pythagorean Theorem

Preceding each section is *Background Information, Management, Procedure,* and *Discussion.*

Student material generally comes in two forms. The first form is for the students for whom an open-ended exploration is more appropriate. At the top of the form are some simple instructions and the situation to be explored. The rest of the sheet contains pictures of Geoboards on which students record their solutions. On the second form, the solutions have been recorded on the Geoboards and students are to determine and record the measurements of each and use their record to develop a generalization of their measurement procedure in the form of a formula. The most benefit comes when students use the first form to develop greater intuitive understandings of perimeter and area and to deal with problem-solving situations. The second form is provided for classes requiring more structure or are limited in time and cannot allow for as much exploration.

Students should have a variety of experiences with each type of geometric figure. Geoboards provide one of those experiences. Between Geoboard sections have students explore the same type of regions with paper cutting, geo-sticks, algebra blocks, or puzzles. As students become experienced with each type of region, they should be asked to work backwards to find missing components.

Various management strategies work well for the exploration of the different types of geometric regions. The rectangle section works well with students working individually. The parallelogram section works well in small groups of two or four students. Whole class involvement in the triangle section with its intense discussion lends itself to the development of a class solution set that can be recorded on the bulletin board. As students gain expertise in developing formulas from Geoboard experiences, the trapezoid activity becomes a small group assessment on the understanding of area and formulas.

* Reprinted with permission from *Principles and Standards for School Mathematics,* 2000 by the National Council of Teachers of Mathematics. All rights reserved.

Class Records

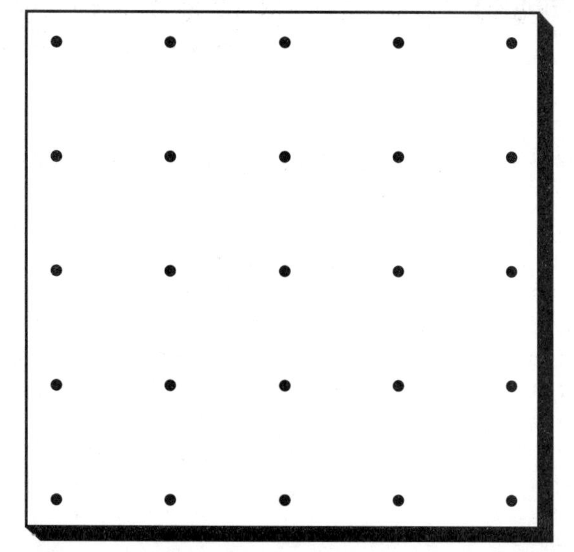

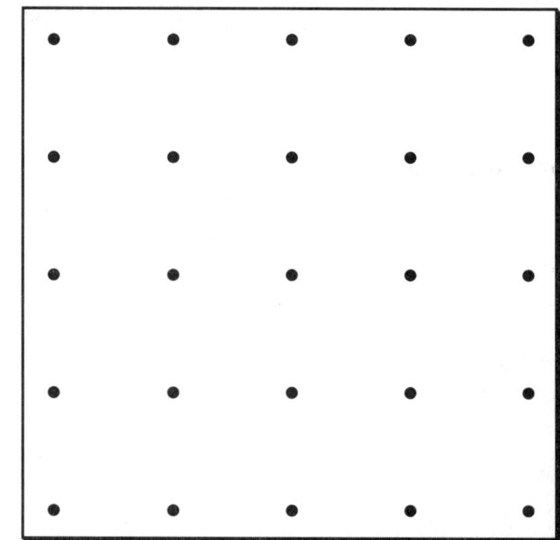

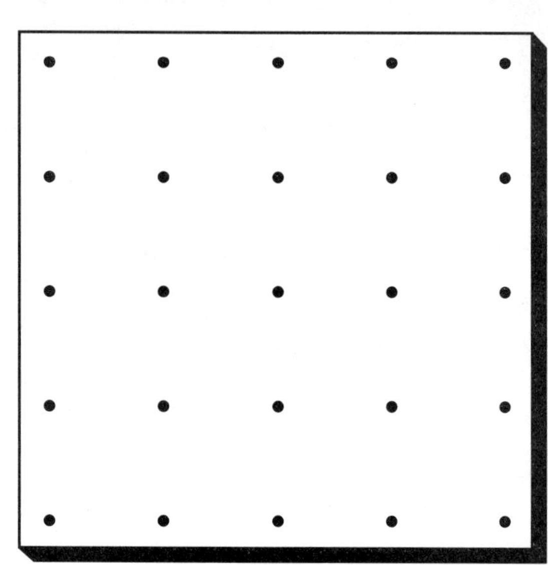

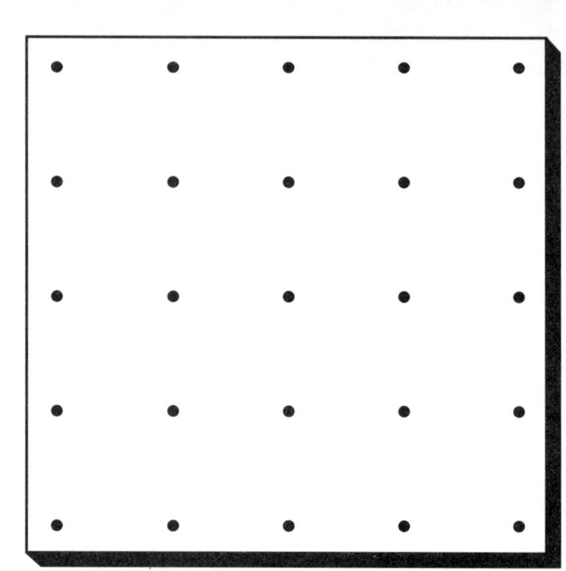

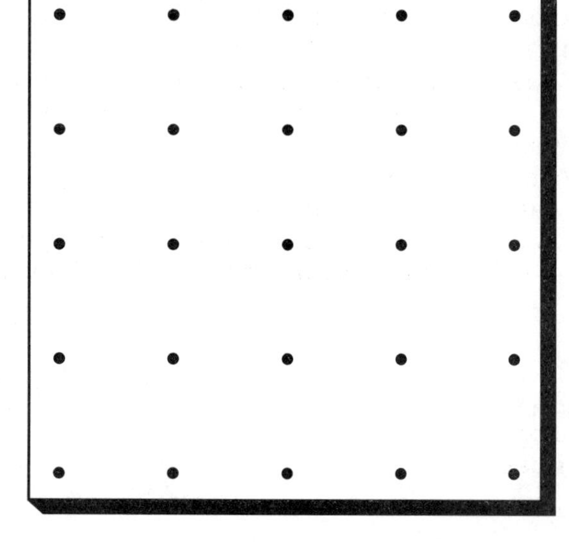

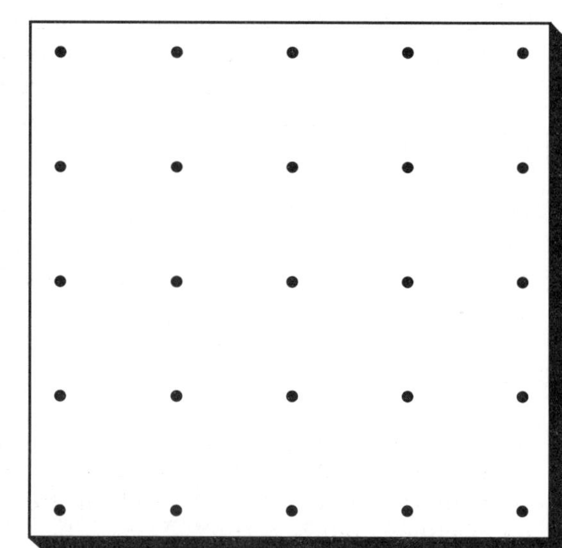

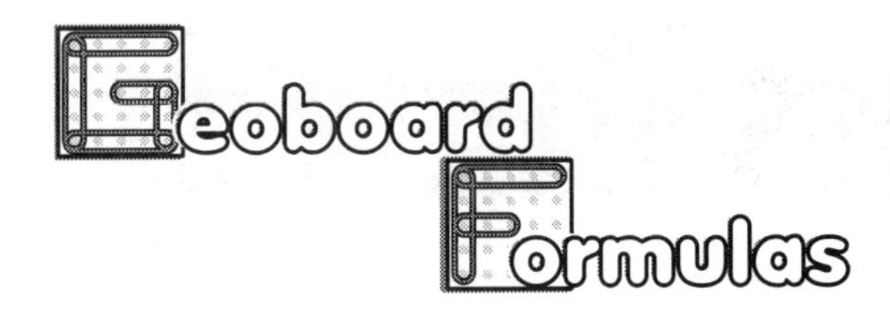

Background Information

The first activity, *Rectangles—A* or *Rectangles—A.1*, has students examine simple closed curves made with right angles. Although it does not look only at rectangles, it has students look at a variety of regions with equal perimeters and varying areas. It provides a simple experience of determining area by counting squares in sub-regions. The concept of perimeter is reinforced as students check the perimeters of many regions. Students can gain insight into perimeter if they identify and observe all the regions that are at least four units wide and high at some point. These regions all have the same perimeter but vary in area. By asking for an explanation as to how these regions can have the same perimeters, students will recognize that parts of the perimeter of the original square have been cut and translated. All the components of the original remain, they have only been slid up or down, or right or left.

The second investigation, *Rectangles—B* or *Rectangles—B.1*, has students find all possible rectangular regions on a Geoboard with sides parallel to the edges of the Geoboard. This provides an excellent experience for students to develop an organized list. Be aware that a discussion about why a square is a rectangle may need to take place. The ten rectangles that can be constructed provide enough practice to develop formulas for the area and perimeter of rectangles.

Management

1. If an adequate supply of materials is available, have students work individually. These introductory activities are simple enough that students can find all the solutions on their own and they get more practice.
2. Determine if the class will work on the open-ended record sheets (A) or the record sheets with the drawn solutions (A.1). These are easy enough that most students are successful with little guidance.

Procedure

Rectangles—A or *A.1*

1. Distribute the materials to the students.
2. Direct the students to find and record 12 regions with perimeters of 16 units and all their sides parallel to the edges of the Geoboard.
3. Have them calculate and record the area of each region.

Rectangles—B or *B.1*

1. Distribute the materials to the students.
2. Direct the students to find and record all the different-sized rectangles that can be constructed on a Geoboard so all the sides are parallel to the edges of the Geoboard.
3. Have them calculate and record the area and perimeter of each rectangle.
4. Discuss with the students what patterns they observed.

Discussion

Rectangles—A or *A.1*

1. Identify any regions that are not at least four units high and four units wide at some point. How are all these regions similar? [They all have some part that sticks into the region—concave.]
2. What is true about all the areas of the regions that are four units high and wide at some point? [The areas are different.]
3. How can all these shapes have the same perimeter as the four by four square? [Translation of edges: the total length remains the same, the segments of an edge have moved.]

Rectangles—B or *B.1*

1. How can you be sure you have all the possible rectangles? [make an organized list]
2. How could you generalize a way to find the perimeter and write it as a formula? [l+w+l+w=P, 2l+2w=P, 2(l+w)=P]
3. How could you generalize a way to find the area and write it as a formula? [l•w=A]

Extension

For Rectangles—B

Is there a pattern to the number of possible rectangles and the size of the Geoboards?

Find 12 regions that have a perimeter of 16 units and have all their sides parallel to the edges of the Geoboard.

- Determine and record the area of each region.

These 12 regions have a perimeter of 16 and all their sides are parallel to the edges of the Geoboard.

- Determine and record the area of each region.

How many different-sized rectangles can be constructed on a Geoboard so all the sides are parallel to the edges of the Geoboard?

- Determine and record the perimeter and area of each rectangle.
- Generalize a pattern and write a formula that tells how to find the perimeter and area of a rectangle.

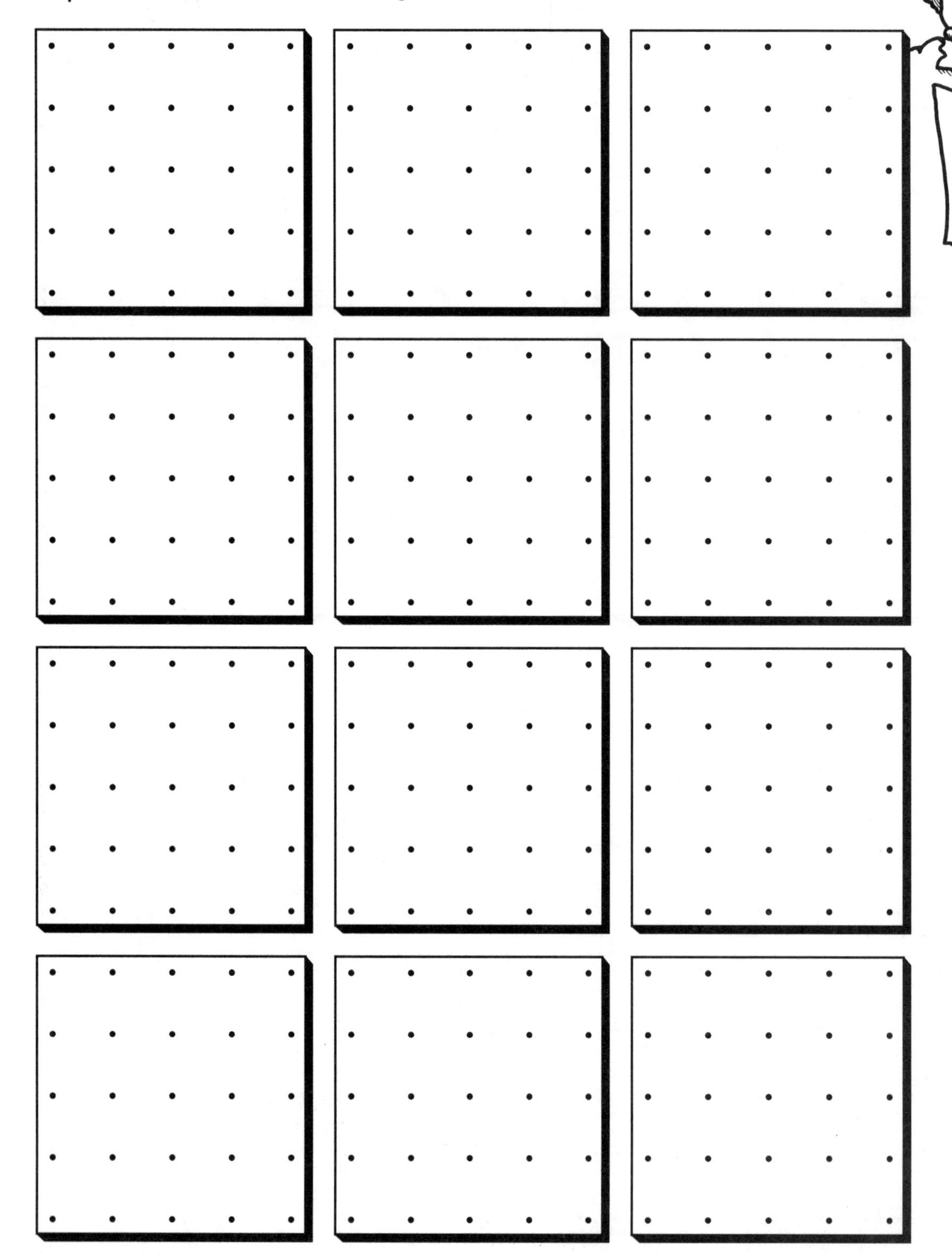

Geoboard Formulas Rectangles-B.1

How many different-sized rectangles can be constructed on a Geoboard so all the sides are parallel to the edges of the Geoboard?

- Determine and record the perimeter and area of each rectangle.
- Generalize a pattern and write a formula that tells how to find the perimeter and area of a rectangle.

Background Information

Any parallelogram can be turned into a rectangle with the same base, height, and area. The Geoboard encourages this observation by providing lines with which to dissect the parallelogram. Consider the parallelograms on the structured record sheet. When all the vertical lines made by pegs are considered, the resulting pieces can be moved to make the basic rectangle that has the same base and height as the parallelogram. The resulting number sentence is just the base times the height of the parallelogram since it is the same length and width of its basic rectangle.

Some students will make a rectangle that surrounds the parallelogram. They will determine the area of the two triangles of the rectangle that are not part of the parallelogram and subtract the triangles' areas from the rectangle. This method does not easily translate into a number sentence but can provide a valuable insight. If all the parallelograms of equal area are grouped together, as they are in the columns of the first page of the structured record sheet, students will observe that their tops have been skewed. The top section of the rubber band has just been slid sideways. It still has the same size top and bottom and height. It has just been "kicked" sideways. Students are quick to observe that you can take any parallelogram and "kick" it back into a rectangle that has the base and height of the parallelogram.

Management

1. The number of parallelograms that can be generated makes this a good investigation to be done in small groups of two to four students.
2. Determine if the class will work on the open-ended record sheets or the record sheets with the drawn solutions.
3. All the solutions are recorded on the structured record sheet except the four by four square.

Procedure

1. Distribute the materials to the students.
2. Direct the students to find and record all the different-sized parallelograms with at least one pair of edges parallel to the edges of the Geoboard.
3. Have them calculate and record the area of each parallelogram. Encourage them to consider a variety of calculation methods by discussing their procedures with each other.
4. Ask the students to sort the parallelograms into groups of equal areas and observe the patterns that exist within each group.
5. Discuss with the students what patterns they observed in their investigation.

Discussion

1. How are the parallelograms with the same areas similar? [bases are the same length, heights are the same]
2. How could you prove that one parallelogram is equal in area to another without counting? [cut them on lines perpendicular to their parallel edges and reassemble them as congruent rectangles]
3. How could you generalize the patterns you found as a formula to determine the area of a parallelogram? [base • height = Area]

How many different-sized parallelograms with at least one pair of edges parallel to the sides of the Geoboard can be made?

- Determine and record the area of each parallelogram.
- Generalize a pattern and write a formula that tells how to find the area of a parallelogram.

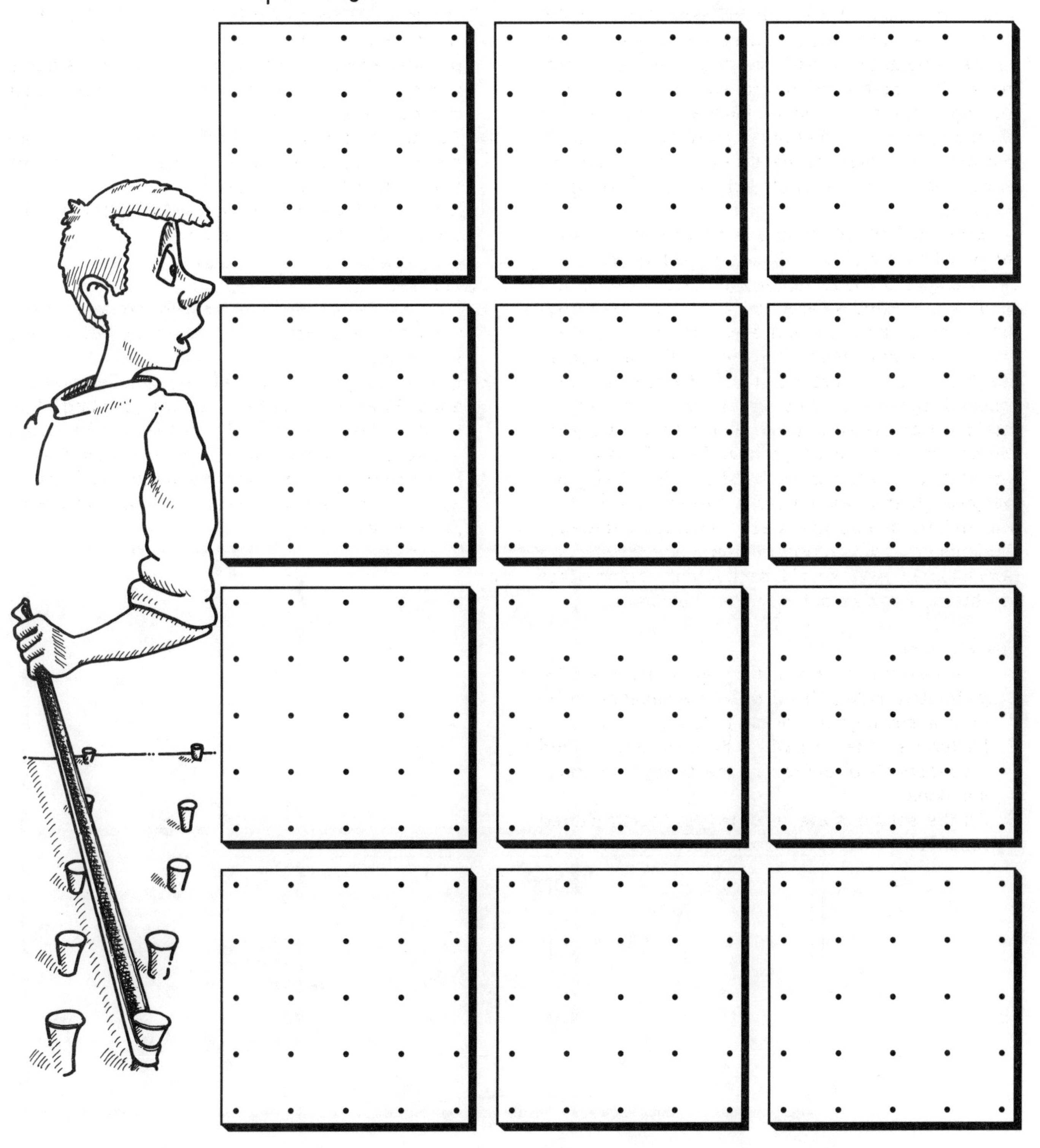

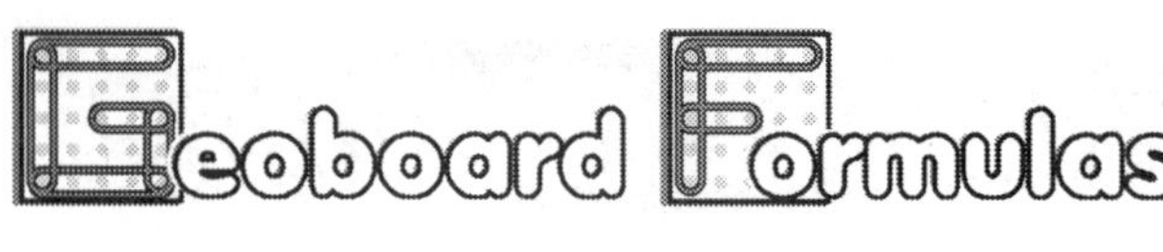

- Determine and record the area of each parallelogram.
- Generalize a pattern and write a formula that tells how to find the area of a parallelogram.

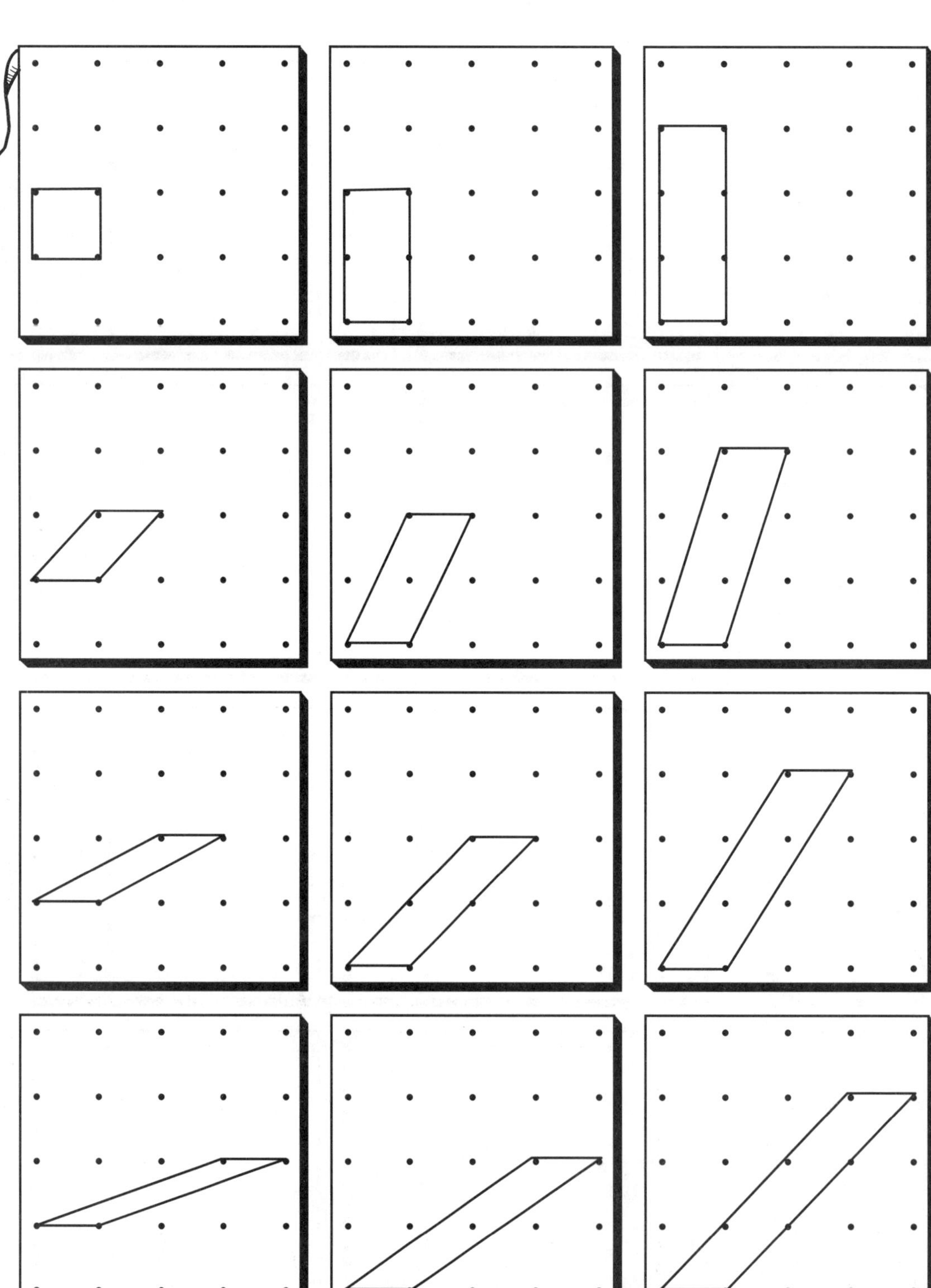

- Determine and record the area of each parallelogram.
- Generalize a pattern and write a formula that tells how to find the area of a parallelogram.

- Determine and record the area of each parallelogram.
- Generalize a pattern and write a formula that tells how to find the area of a parallelogram.

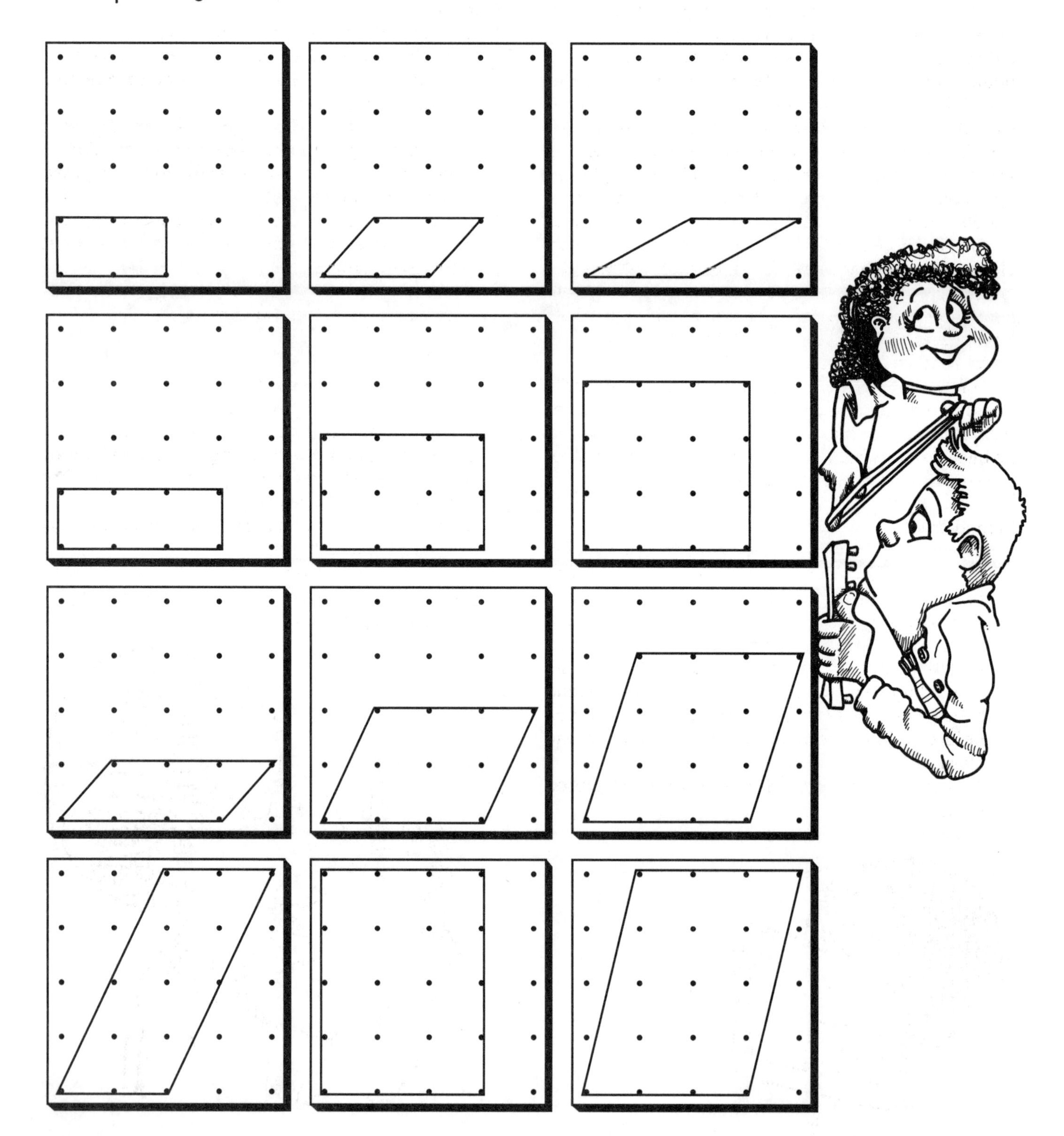

Background Information

Students working with right triangles are quick to recognize that a right triangle is half of a rectangle. Using this insight, students often surround obtuse and acute triangles with a rectangle and subtract the two right triangles that are not part of the original triangle. This method will not easily translate into a formula. However, if students are encouraged to sort the triangles into groups of equal area, as in the columns of the first page of the structured record sheets, they will be able observe the similarities. They will recognize that the top vertex has been skewed, the rubber band has moved off to the side. Each triangle can be transformed into a right triangle simply by moving the top vertex parallel to the base. So any triangle is half the area of the rectangle formed by its base and height.

A few students may recognize that if you replicate any of the triangles and put the two together, you get a parallelogram. Since the area of the parallelogram is its base times its height, the area of a triangle is half of the parallelogram that has its base and height.

Management

1. Determine if the class will work on the open-ended record sheets or the record sheets with the drawn solutions.
2. The number of triangles that can be generated in an open-ended format makes this a good investigation to be done as a class with students submitting solutions to a central location (bulletin board, table) where they can be sorted into groups before being recorded by each student in the class.
3. All the solutions are recorded on the structured record sheet.

Procedure

1. Distribute the materials to the students.
2. Have the students find and record all the different-sized triangles with at least one edge parallel to a side of the Geoboard.
3. Direct them to calculate and record the area of each triangle. Encourage them to consider a variety of calculation methods by discussing their procedures with each other.
4. Have the students sort the triangles into groups of equal areas and observe what patterns exist within each group.
5. Discuss with the students what patterns they observed in their investigation.

Discussion

1. How are the triangles with the same area similar? [bases the same length, heights the same]
2. How could any triangle in a group be changed into a right triangle without changing its area? [Slide a vertex parallel to the opposite base until the triangle is a right triangle.]
3. What shape can you make with any two congruent triangles? [parallelogram]
4. How could you generalize the patterns you found as a formula to determine the area of a triangle? [(base • height)÷2 = Area]

How many different-sized triangles with at least one edge parallel to a side of the Geoboard can be made?

- Determine and record the area of each triangle.
- Generalize a pattern and write a formula that tells how to find the area of a triangle.

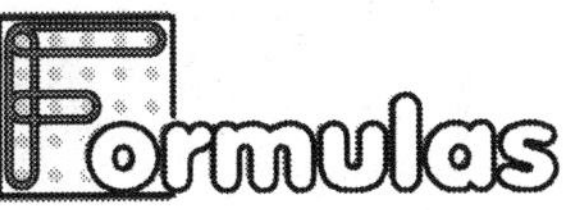

- Determine and record the area of each triangle.
- Generalize a pattern and write a formula that tells how to find the area of a triangle.

Geoboard Formulas

- Determine and record the area of each triangle.
- Generalize a pattern and write a formula that tells how to find the area of a triangle.

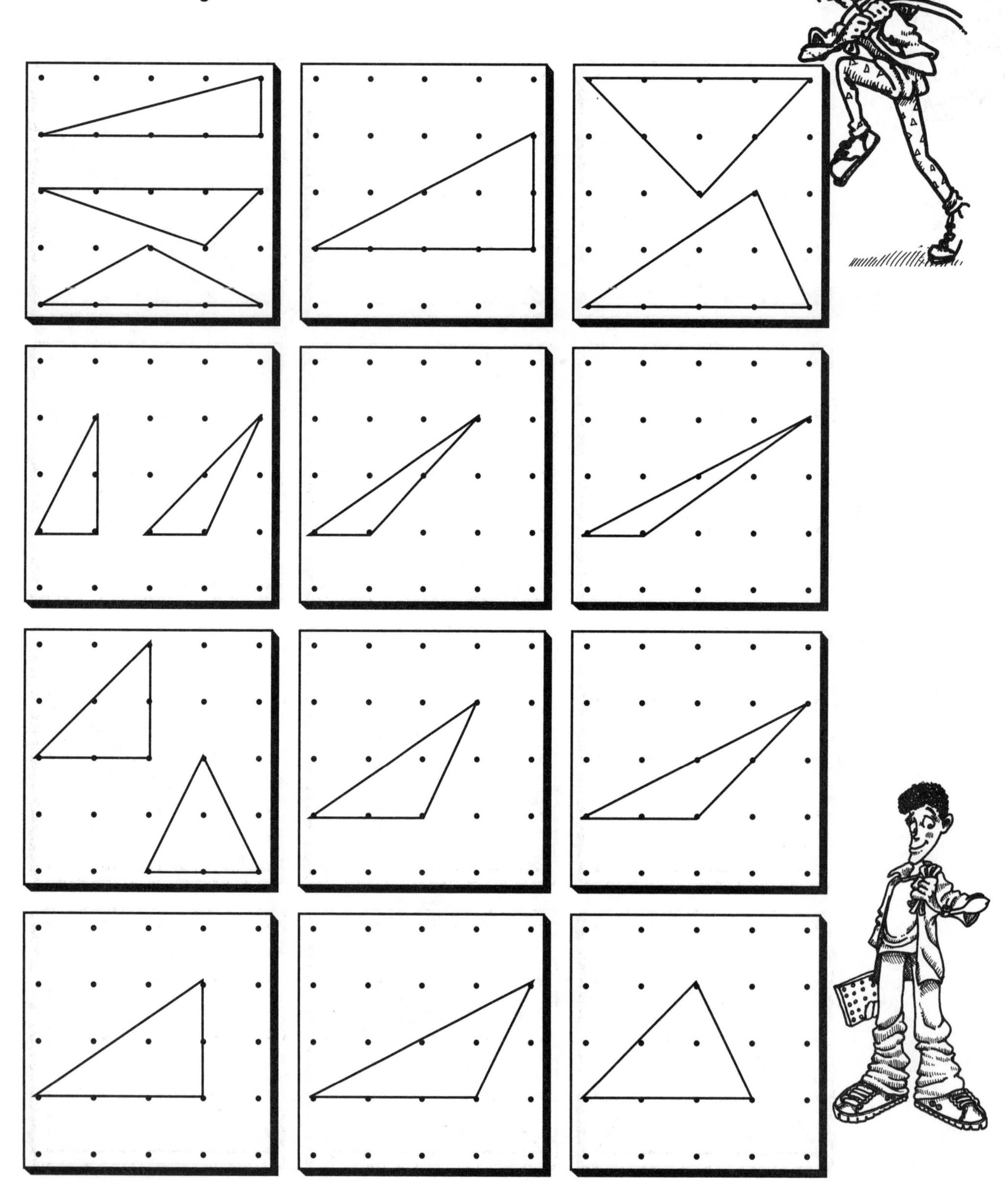

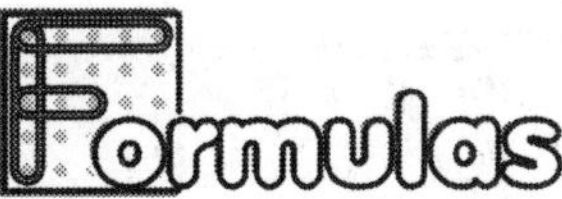

- Determine and record the area of each triangle.
- Generalize a pattern and write a formula that tells how to find the area of a triangle.

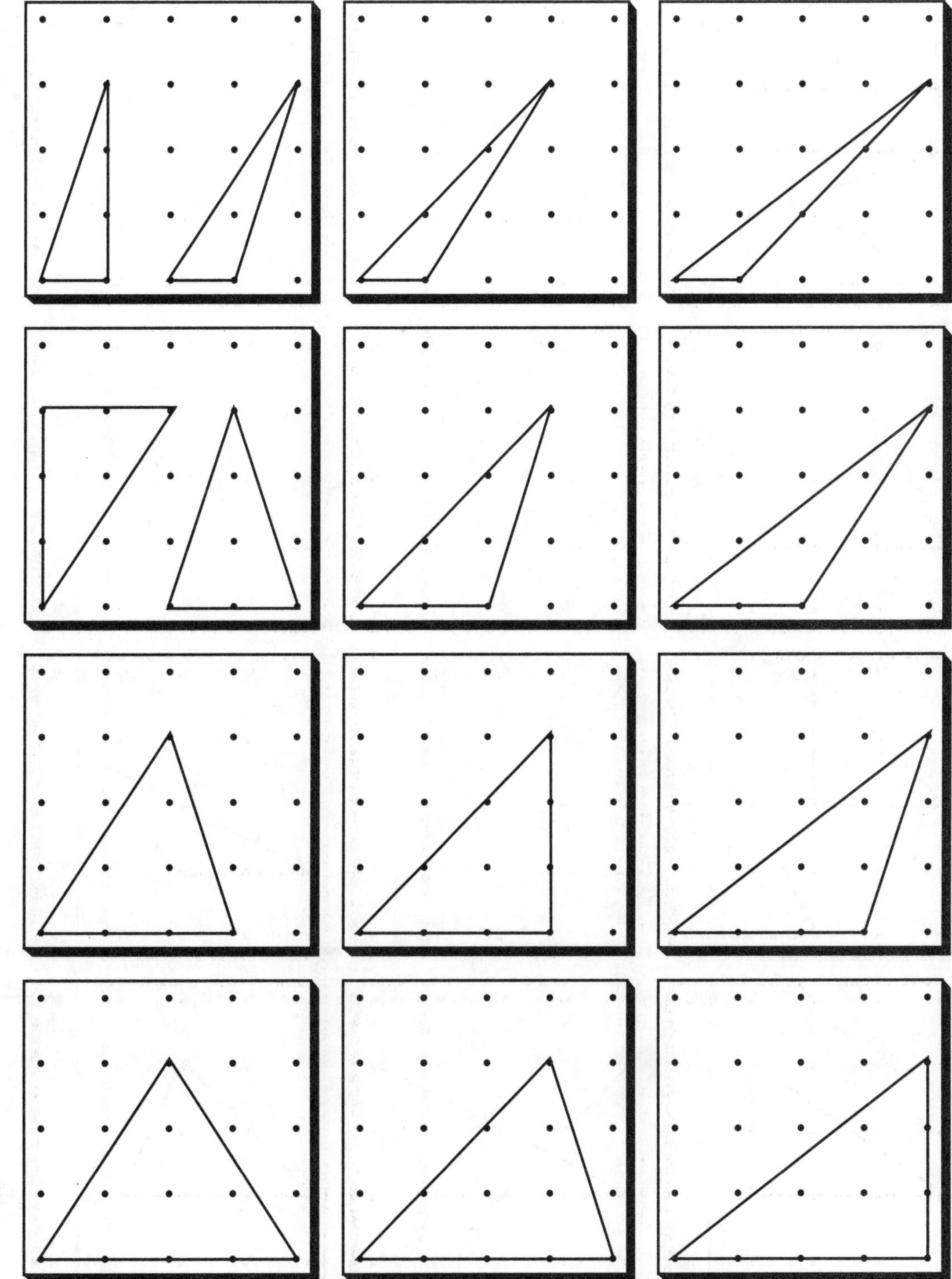

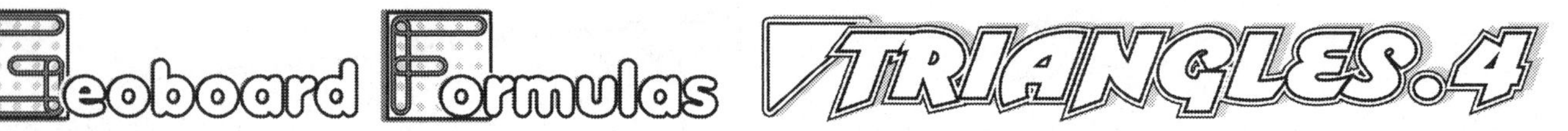

Geoboard Formulas — Triangles.4

- Determine and record the area of each triangle.
- Generalize a pattern and write a formula that tells how to find the area of a triangle.

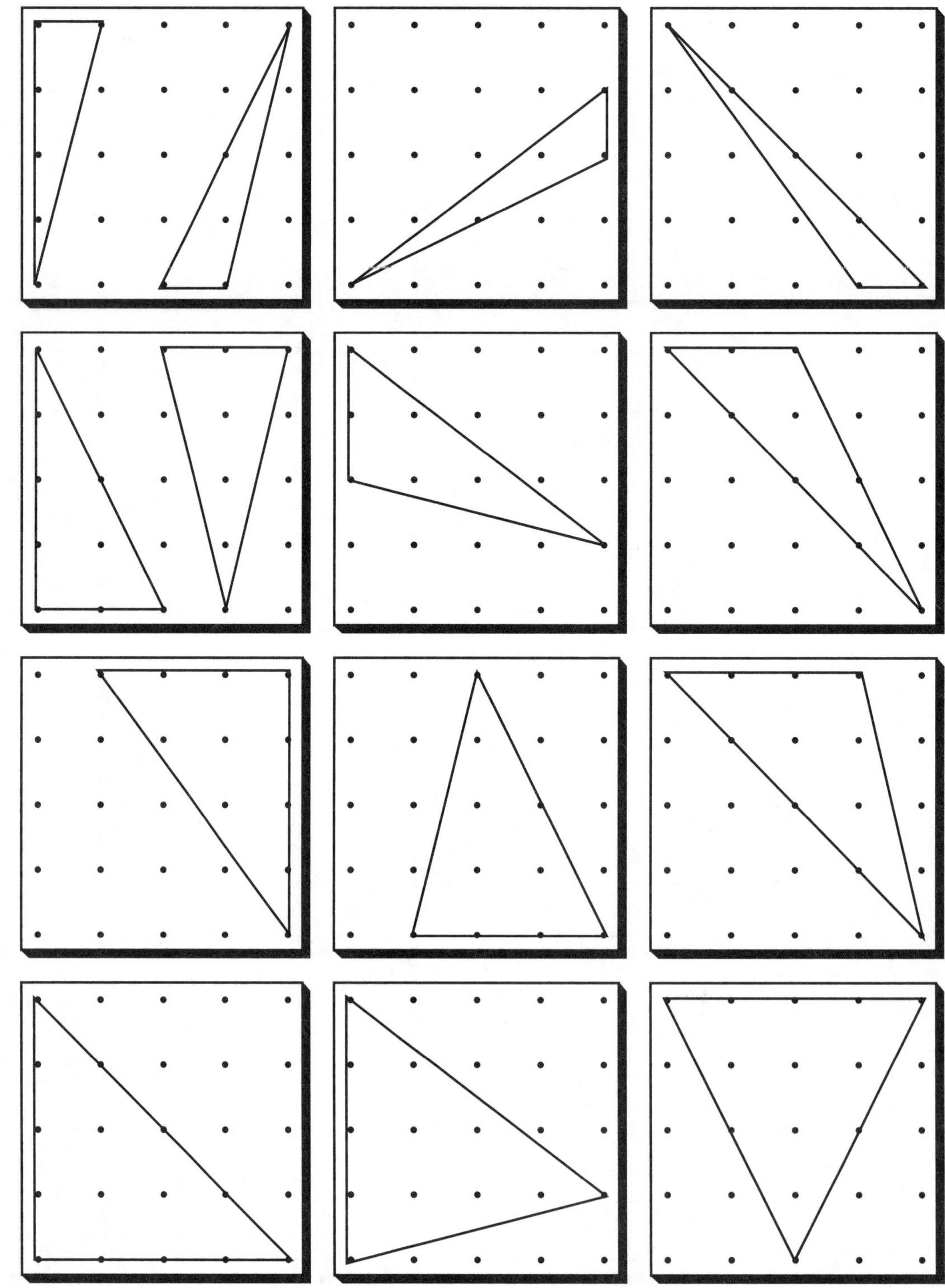

TRAPEZOIDS

Background Information

Developing formulas for trapezoids is more difficult for students than other geometric figures so having more experiences prior to dealing with trapezoids and then developing a large set of trapezoids to study are beneficial.

As students work with trapezoids, many will recognize that all of them can be dissected into a rectangle and a triangle. By disregarding the rectangle, students can see there are two triangles remaining. They can easily imagine sliding these two together to form one triangle. The area of the rectangle is found by multiplying the shorter base (b_1) by the height ($b_1 \bullet h$). The base of the triangle is what is left of the longer base (b_2) when the shorter base is taken away (b_2-b_1). The area of the triangle is calculated using the formula $(((b_2-b_1)\bullet h)\div 2)$. So the formula for the area of a trapezoid from this perspective is $A=(b_1 \bullet h)+(((b_{2.}b_1)\bullet h)\div 2)$.

Some students will see that every trapezoid can be cut into two triangles with the bases of the trapezoid as bases to the triangles. Using the triangle formula, students can generate the following variation of the trapezoid formula: $A = (b_1 \bullet h \div 2) + (b_2 \bullet h \div 2)$.

Visually students do not tend to develop the variation traditionally given, $A = (b_2+b_1)/2 \bullet h$.

Management

1. Because of the number of trapezoids possible on a Geoboard, it is recommended students work in small groups of two to four. This provides an excellent time to assess their understanding of calculating area and their ability to change visual patterns into formulas.
2. Students will have more success on open-ended record sheets if they record all the possible trapezoids one unit high on one page, all the trapezoids two units high on another page, and all the trapezoids three and four units high on a third and fourth page.

Procedure

1. Distribute the materials to the students.
2. Direct the students to find and record all the trapezoids that can be made on a Geoboard where the parallel edges of the trapezoid are parallel to the sides of the Geoboard.
3. Have them calculate and record the areas of the trapezoids. Encourage them to consider a variety of calculation methods by discussing their procedures with each other.
4. Discuss with the students what patterns they observed in their investigation.

Discussion

1. How did you calculate the area of the trapezoids? (Refer to *Background Information*.)
2. Into what shapes can you cut all the trapezoids? [2 triangles, 1 rectangle and a triangle]
3. How could you generalize the patterns you found as a formula to determine the area of a triangle? [$A = (b_1 \bullet h \div 2) + (b_2 \bullet h \div 2)$, $A=(b_1 \bullet h)+(((b_{2.}b_1)\bullet h)\div 2)$]

How many trapezoids can be be made where the parallel edges are parallel to the sides of the Geoboard? (Hint: Make all 1 unit high, then 2 units, 3 units, and 4 units.)

- Determine and record the area of each trapezoid.
- Generalize a pattern and write a formula that tells how to find the area of a trapezoid.

- Determine and record the area of each trapezoid.
- Generalize a pattern and write a formula that tells how to find the area of a trapezoid.

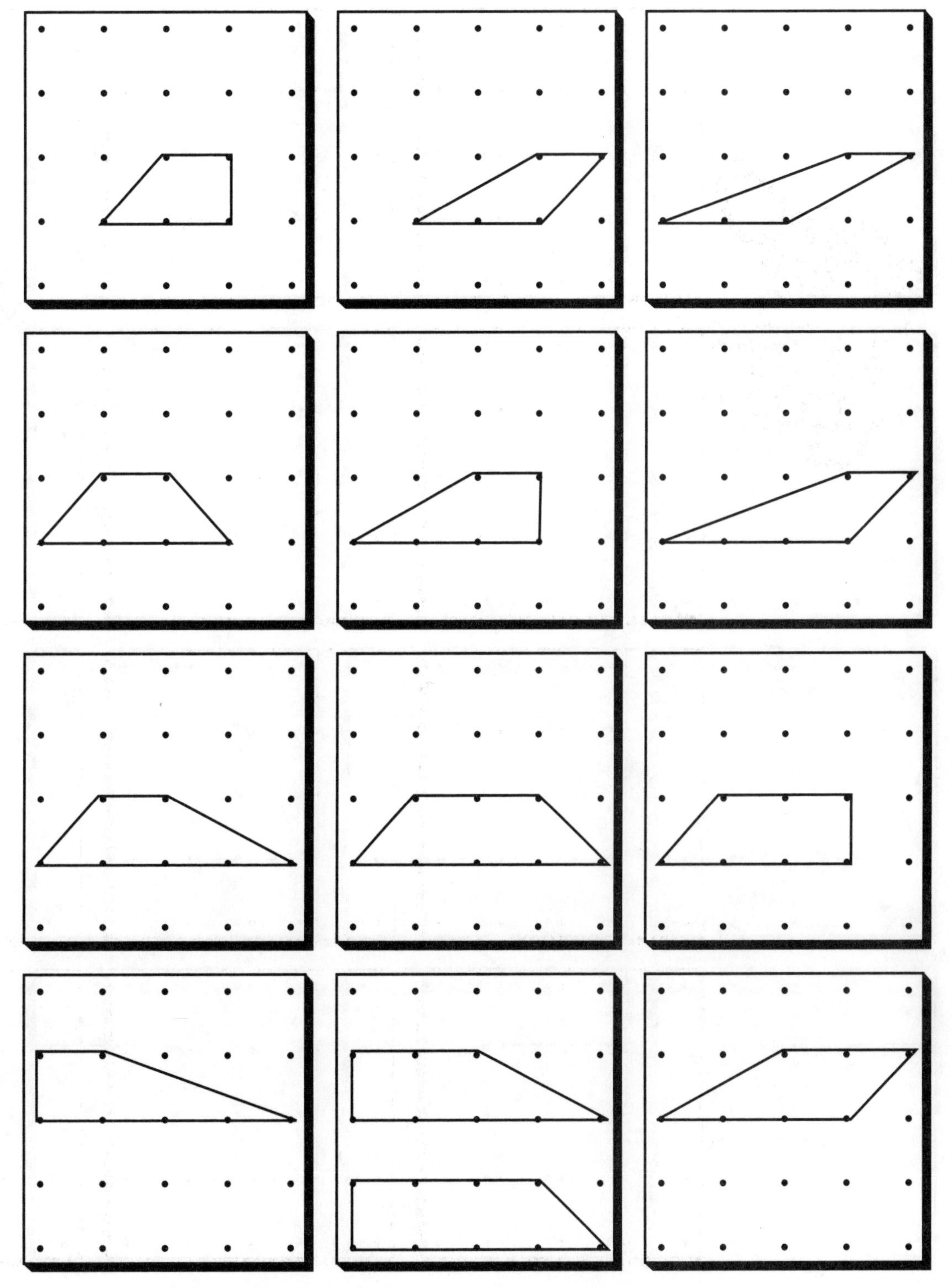

- Determine and record the area of each trapezoid.
- Generalize a pattern and write a formula that tells how to find the area of a trapezoid.

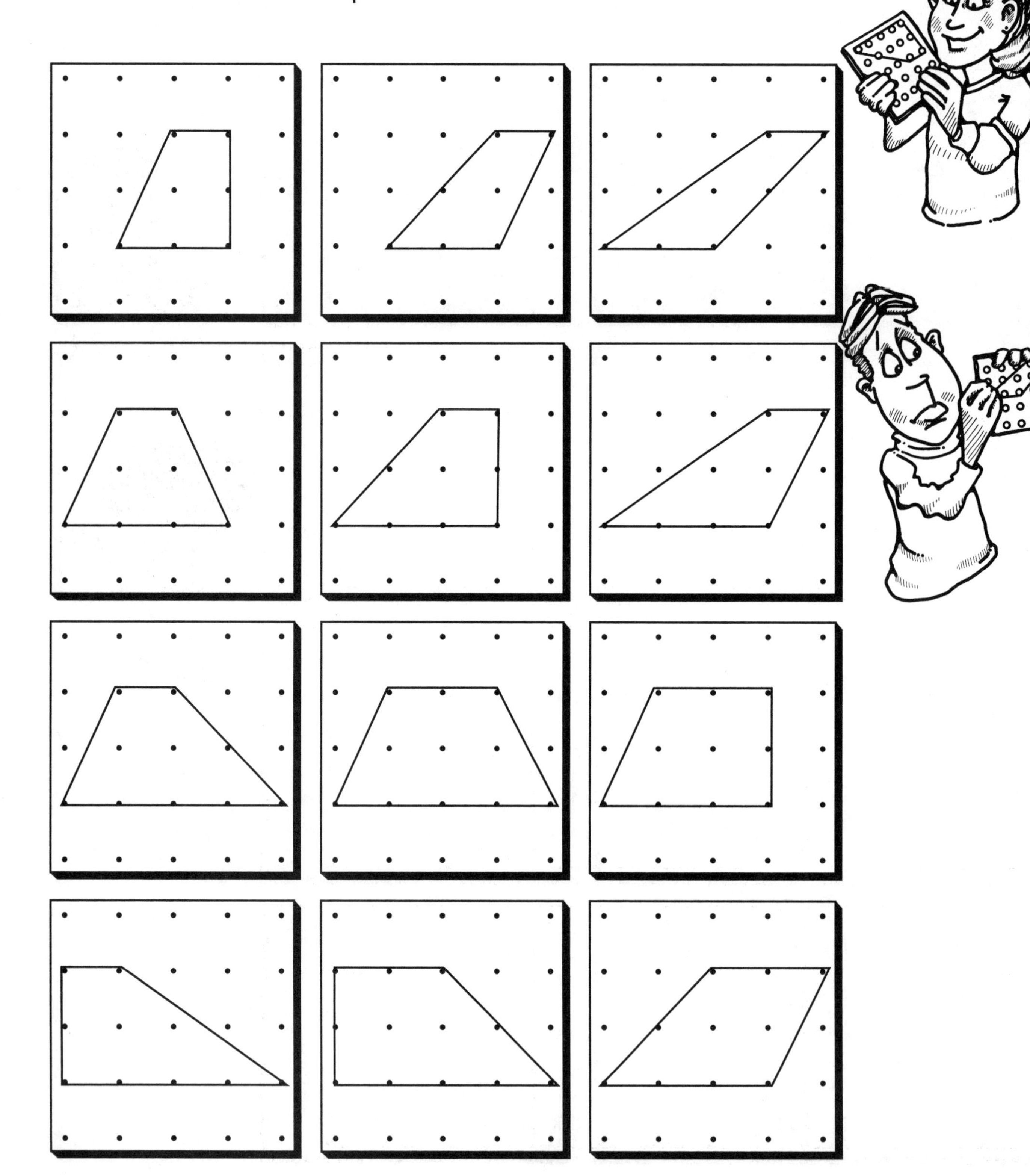

- Determine and record the area of each trapezoid.
- Generalize a pattern and write a formula that tells how to find the area of a trapezoid.

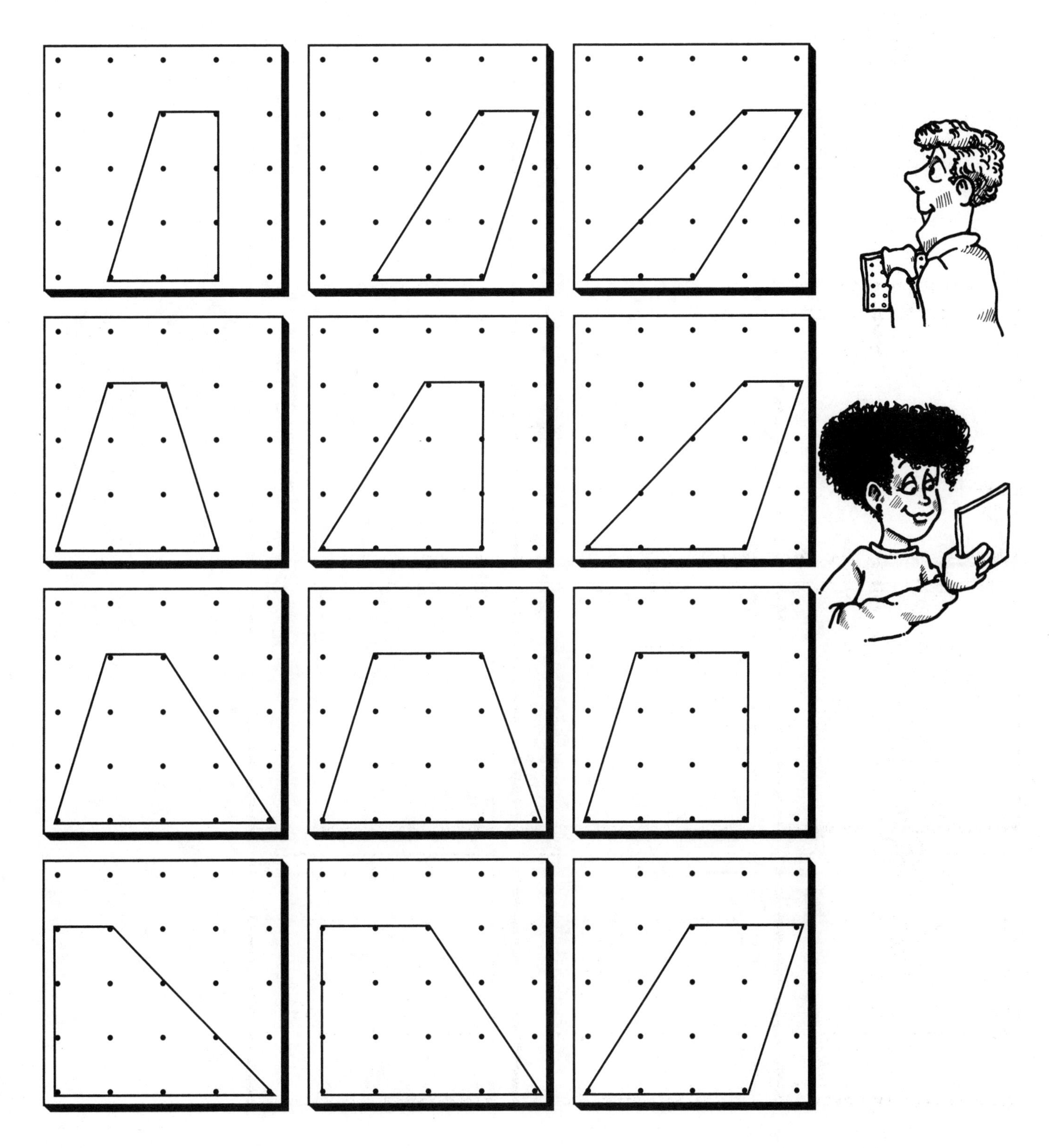

- Determine and record the area of each trapezoid.
- Generalize a pattern and write a formula that tells how to find the area of a trapezoid.

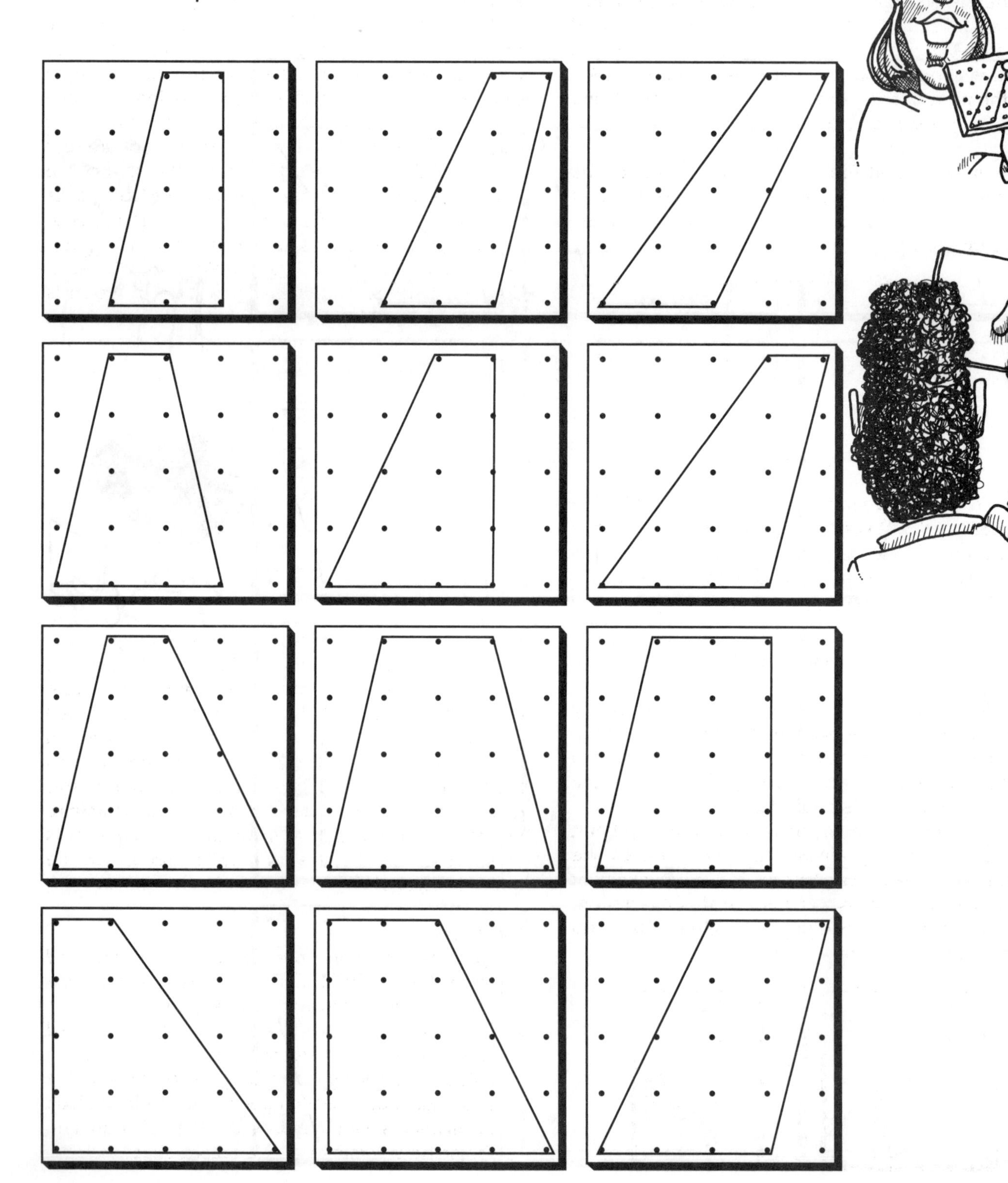

PYTHAGORAS

Background Information

To deal with the Pythagorean theorem, students must have a good understanding of square numbers and square roots. By building all the possible squares on the Geoboard, students can relate a visual understanding to the numeric equivalent. The squares that have sides parallel to the edges of the Geoboard form the perfect squares and the idea that $2^2 = 4$ and $\sqrt{(4)} = 2$, $3^2 = 9$ and $\sqrt{(9)} = 3$, and $4^2 = 16$ and $\sqrt{(16)} = 4$ develops clearly.

The squares that form at angles diagonal to the Geoboard edges extend this idea to a more abstract level. The area of one of the squares can be counted as 2 square units.

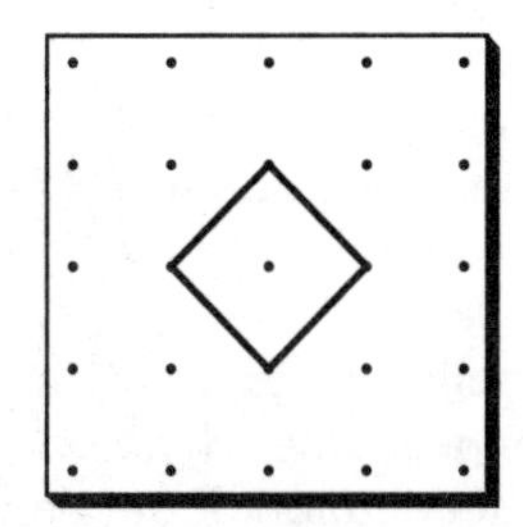

Since the edge of a square is the square root length of its area, the length of that square is 1.41421.... It is a stretch for children to accept a radical as a solution when they are used to dealing with numbers. As students work the other problems ($\sqrt{(5)}$ 2.23606..., $\sqrt{(8)}$ 2.82842..., $\sqrt{(10)}$ 3.16227...) the relationship of square and square root is reinforced in a visual way.

Students will discover the Pythagorean theorem by finding the areas of the squares formed on the three sides of right triangles. Since the legs are aligned with the peg grid, drawing and counting the areas of the squares on the legs is straight-forward.

The square of the hypotenuse is more difficult because it is not aligned with the grid. To draw the squares, one needs to consider the relationship of the endpoints of the hypotenuse. In this illustration, the upper endpoint is one peg right and two pegs up from the lower endpoint.

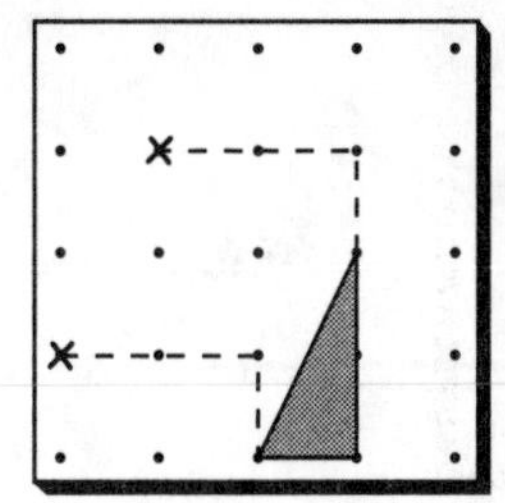

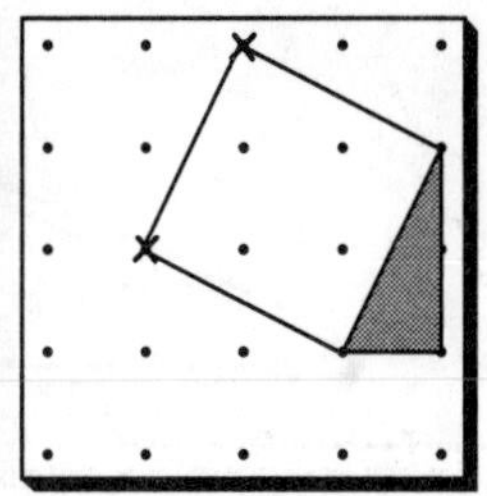

The other point can be found by reversing the distances and the lateral direction so the point is one peg up and two pegs to the left of the hypotenuse's endpoint.

Students tend to determine the area of the misaligned squares in one of two ways. One is to divide it into component parts of squares and right triangles. On the interior of each misaligned square is a smaller square of complete squares made by the peg grid. Rotated around the interior square are four congruent right triangles. By adding the area of the interior square and the four triangles, the area of the misaligned square is obtained. The second way is to inscribe the misaligned square within a larger square aligned with the peg grid. By subtracting the area of the four congruent right triangles (one of which is the original triangle) needed to complete the aligned square from the total area of the aligned square gives the area of the misaligned square.

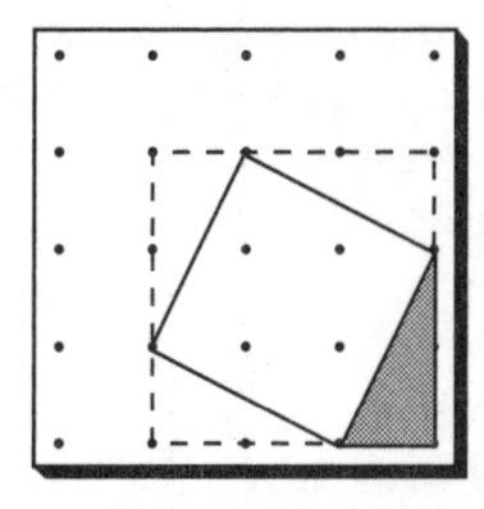

When the area of all three squares on several of right triangles have been recorded, students will recognize that the sum of the areas of the squares of the two smaller legs equals the area of the square on the hypotenuse. If the smaller legs are labeled A and B and the hypotenuse is labeled C, one can say the A square and the B square are the same as the C square ($A^2+B^2=C^2$). As students visualize this generalization in applied settings, they will be able to use their understanding of squares and square roots to approximate the answers while reinforcing the understanding of the relationship in their minds.

Management

1. Each record page should be completed and understood before going to the next one so that the prerequisite understanding is acquired.
2. Determine if the class will work on the open-ended record sheets or the record sheets with the drawn solutions. Drawing the misaligned squares on the Pythagoras open-ended page tends to be difficult for students and requires at least 30 minutes. Providing students with the record sheet with the

squares drawn can provide a significant savings of time.

3. This series can be done individually or in small groups.

Procedure

Part One—Squares

1. Distribute the materials to the students.
2. Have the students find and record all the different-sized squares that can be made on a Geoboard.
3. Direct them to calculate and record the area of each square.
4. Discuss with the students the different methods they used to determine the areas of the squares.
5. Make certain that the students recognize the square to square root relationship in each of the squares.

Part Two—Pythagoras

1. Direct the students to construct squares on all three sides of the triangles on the record sheet.
2. Have them calculate and record the area of each square.
3. Discuss with the students the different methods they used to determine the areas of the squares.
4. Make certain they recognize the patterns of the areas of the squares of each of the triangles and develop a generalization equivalent to the Pythagorean theorem.

Discussion

Part One—Squares

Consider the squares with sides parallel to edges of the Geoboard.

1. What number sentence can you write for each square that relates its sides to its area? [$1 \bullet 1 = 1^2 = 1$, $2 \bullet 2 = 2^2 = 4$, $3 \bullet 3 = 3^2 = 9$, $4 \bullet 4 = 4^2 = 16$]
2. This relationship is true for all squares. How would you generalize the pattern if the length of a side is S units long? [$S \bullet S = S^2$]
3. The length of the side of a square is called the square root of the square. Two equations can be written for each square: $S \bullet S = S^2$ and $\sqrt{(5^2)} = S$. Write two equations for each square. [$1^2 = 1$, $\sqrt{(1)} = 1$; $2^2 = 4$, $\sqrt{(4)} = 2$; $3^2 = 9$, $\sqrt{(9)} = 3$; $4^2 = 16$, $\sqrt{(16)} = 4$]

Consider the squares not aligned with the peg grid.

4. How did you count the area of each square? (Refer to *Background Information.)*
5. Write an equation for each square that gives the length of a side from the area. [$\sqrt{(2)}$ 1.41421..., $\sqrt{(5)}$ 2.23606..., $\sqrt{(8)}$ 2.82842..., $\sqrt{(10)}$ 3.16227...]

Part Two—Pythagoras

1. How did you count the area of each square? (Refer to *Background Information.)*
2. What pattern do you notice in the areas of the squares on each triangle? [The area of the two smaller squares equals the area of largest square.]
3. If the two smaller sides (legs) are called A and B, and the longest side (hypotenuse) is called C, write a general equation of the pattern. [$A^2 + B^2 = C^2$]
4. If you know how long the two legs of a right triangle are, how can you determine the length of the hypotenuse? [Square the legs to get the area of their squares, and add the areas of the two small squares to get the area of the long square. Then get the square root of the long square to get the length of its side, the hypotenuse, $\sqrt{(A^2+B^2)} = C$]

How many different-sized squares can you make on a Geoboard?

- Record each square and determine its area.
- Determine what the length of an edge would need to be from the area of each square.

- Determine and record the area of each square.
- Determine what the length of an edge would need to be from the area of each square.

Is there a pattern in the areas of the squares that are made on the sides of right triangles?

- Find the area of the squares that are made on the sides of the triangles.
- What pattern is there in a triangle's squares?
- How can you write the pattern as a formula?

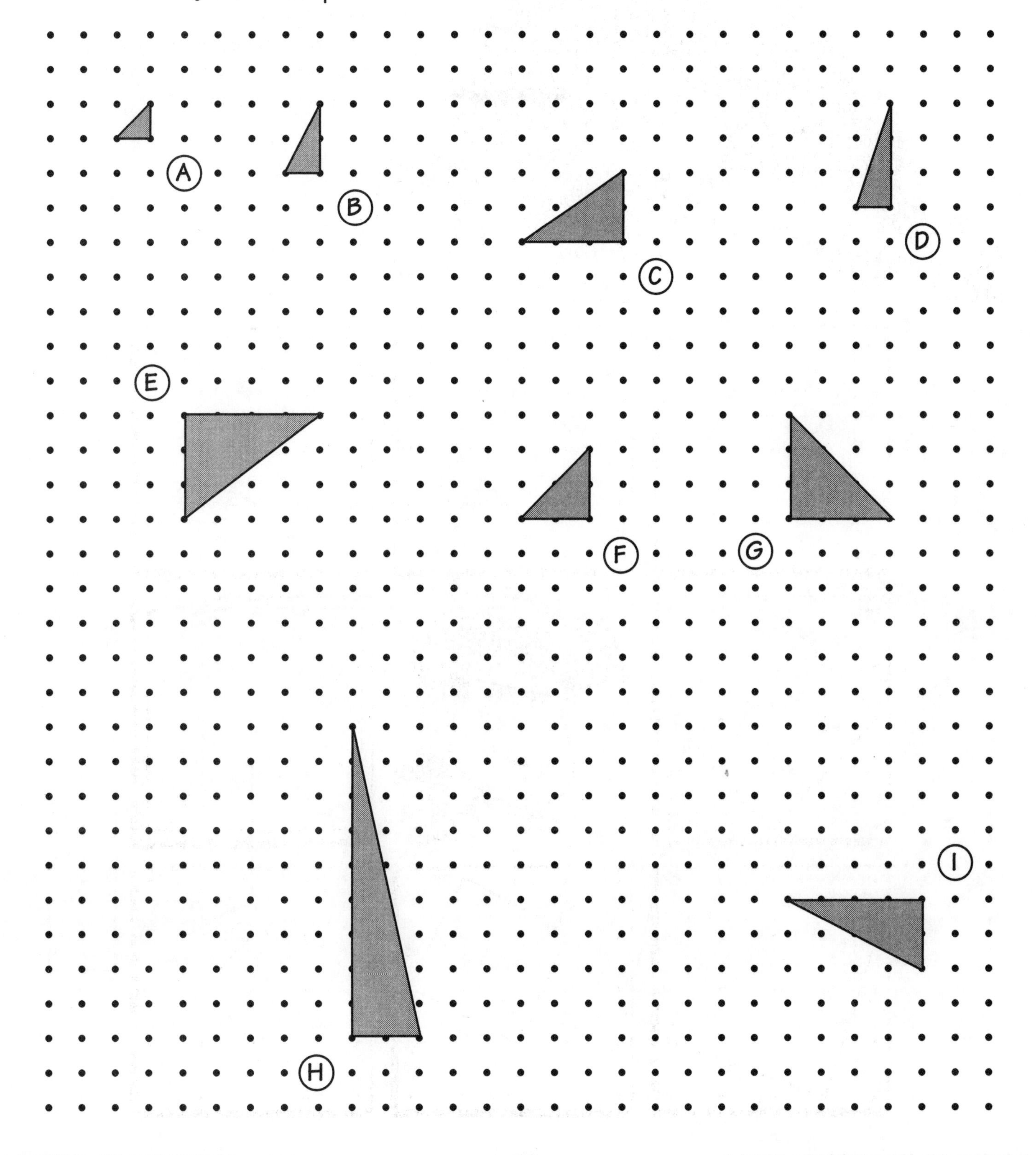

Is there a pattern in the areas of the squares that are made on the sides of right triangles?

- Find the area of the squares that are made on the sides of the triangles.
- What pattern is there in a triangle's squares?
- How can you write the pattern as a formula?

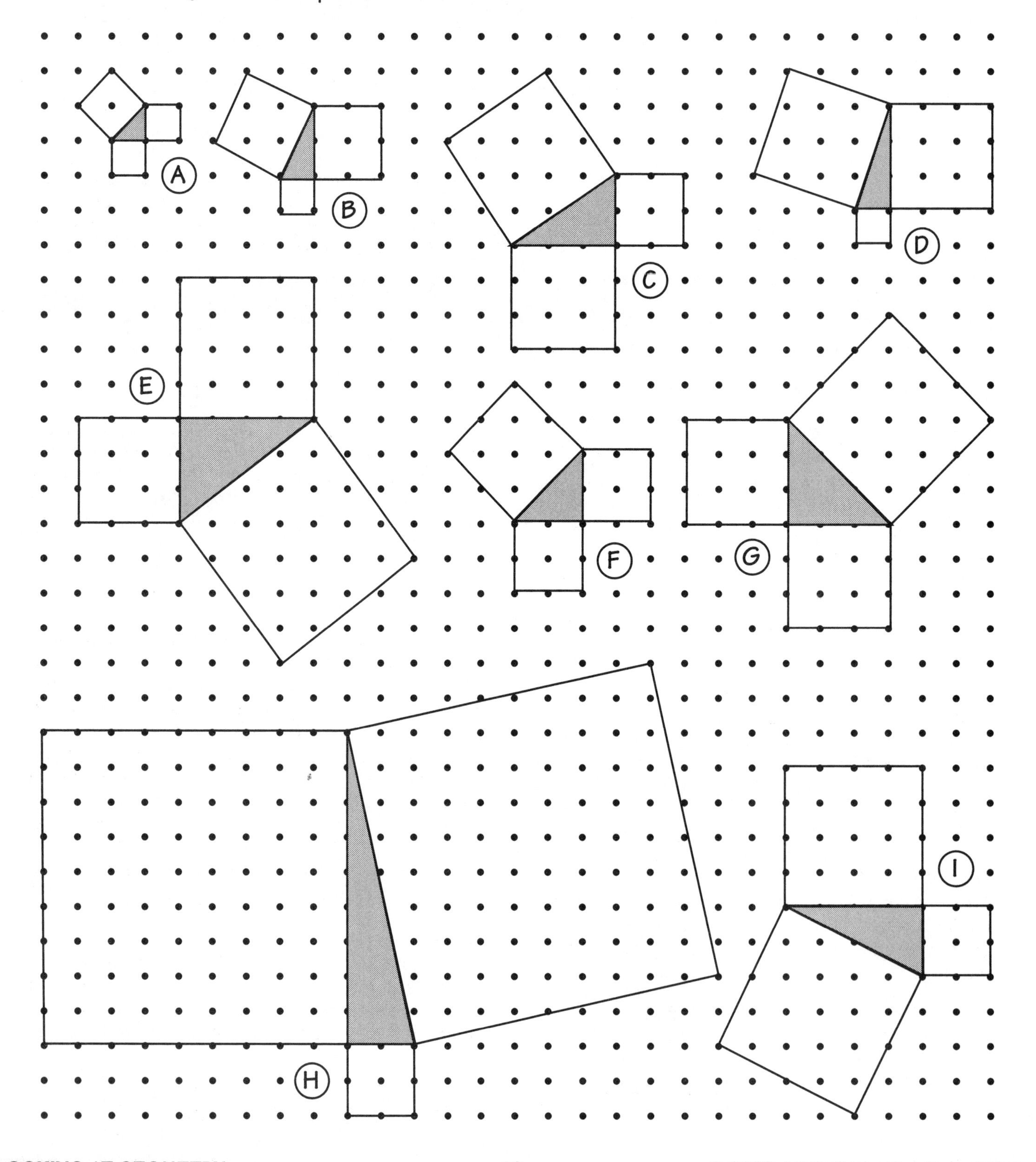

With paper and scissors students can explore relationships between geometrical figures and develop formulas to communicate the area relationships they discover.

Adding machine tape is a convenient source of material for studying polygons. The parallel edges provide a handy reference for making perpendicular or parallel folds and cuts.

Parallel and Perpendicular

When students have been provided with a strip of adding machine tape and scissors, ask them: *How can you make sure that the end of your strip is perpendicular to the strip?* They may have difficulty explaining themselves, but they will try to relate that when part of the strip is folded back on top of itself, the fold line is perpendicular to the edges if the edges line up on top of themselves. This is a natural opportunity to develop the vocabulary of parallel and perpendicular.

Another question might be posed, *How could you get a fold line that is parallel of the edges of your strip?* After some exploration, most students will simply explain that you fold the strip in half with one edge lying on top of the other parallel edge. At this time, you might encourage students to observe the parallel fold line they made and ask them to what edges it is parallel and to what edge it is perpendicular. They will notice that as they folded the side edges on top of each other, they made the fold parallel to the side edges but perpendicular to the end.

Rectangles and Parallelograms

When students understand and can accurately fold and cut parallel and perpendicular lines, they are prepared to consider the relationship of parallelograms to rectangles.

Have students accurately cut rectangles from the adding machine tape and then consider the question, *How many different ways can you make one cut and turn any rectangle into a non-rectangular parallelogram?* It will not take students long to realize that all they need to do is to make a non-perpendicular cut between two parallel edges and then translate one piece.

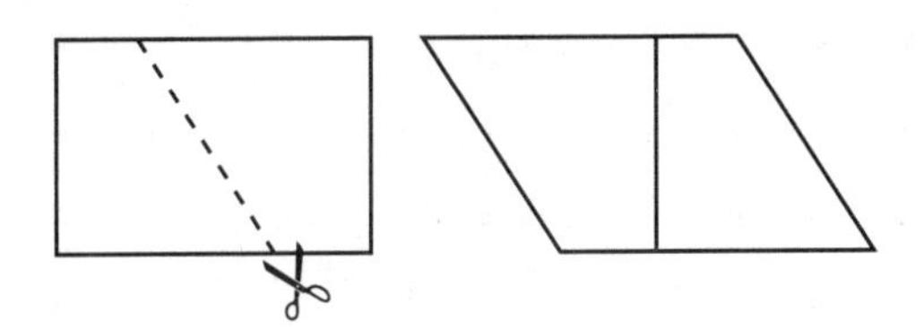

As they compare their solutions to those of others, they will begin to recognize that there are infinite places to cut and get a solution. They will also realize that it is not the position of the cut but its relationship to the edges that is significant.

The relationship of the cut to the edges can be reinforced by starting with parallelograms that are not rectangles. Have students cut a parallelogram by making a loop out of a strip of adding machine tape so the tape is two layers thick and the outside edges line up on top of themselves. By making one cut through both layers across the loop, a parallelogram is formed.

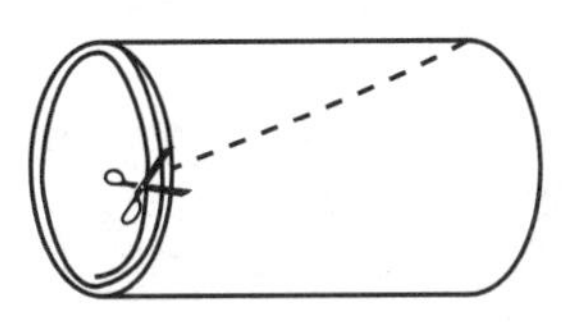

Have the students investigate the question, *How many different places can you cut a parallelogram so the two resulting pieces form a rectangle?* This time students will discover the inverse of their experience with the rectangle. There are infinite solutions, but all of the possible cuts will be made perpendicular to two parallel sides.

When students develop a collection of several rectangles and parallelograms, ask them to develop a formula for determining the area of parallelograms. They should be familiar with the equation for a rectangle (Area = length • width). As students are encouraged to compare the dimensions of the parallelograms with the rectangles they make, they will see that the lengths or bases are equal and the width of the rectangle is the perpendicular cut through the parallelogram called the height or altitude. The equation for the parallelogram (Area = base • height) is the same as the rectangle except for the names that have been substituted to clarify the dimension used from the parallelogram.

Triangles

To study the relationship of triangles to parallelograms, have students cut two congruent triangles. Students can do this by folding a strip of adding machine tape on top of itself with the edges lined up. Using one edge as a side, have them make two intersecting cuts through both layers of paper.

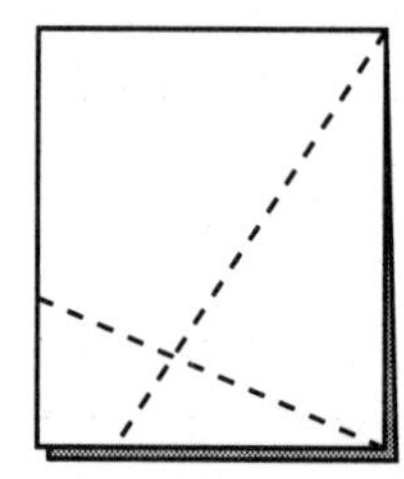

Have them consider this question while looking at the variety of samples made in the class, *Will any two congruent triangles form a parallelogram?* Seeing that any two congruent triangles form a parallelogram, students can use their prior learning to develop the equation: Area = (base • height) ÷ 2 because a triangle is half the area of the parallelogram that has a base and height congruent to the triangle.

Another relationship of dimensions within a triangle can be seen by having the students follow these set of instructions:

1. Cut out two congruent triangles and place one in a position where the vertex is directly above the base. One will not be altered so that it can be used as a comparison.

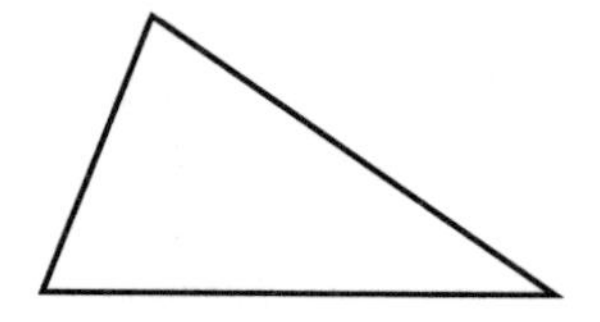
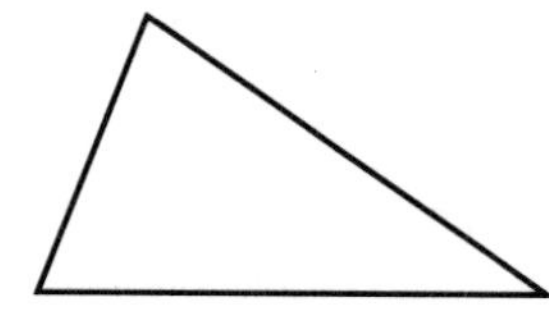

2. Fold the base on top of itself so the fold will be perpendicular to the base. Move the base across itself until the fold goes through the vertex. Crease to form the height line.

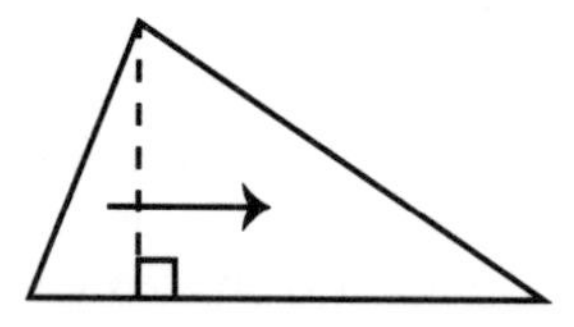

3. Unfold the triangle. Now, fold the top vertex down to the base so it touches the base at the height line. This fold forms a line parallel to the base at half the triangle's height.

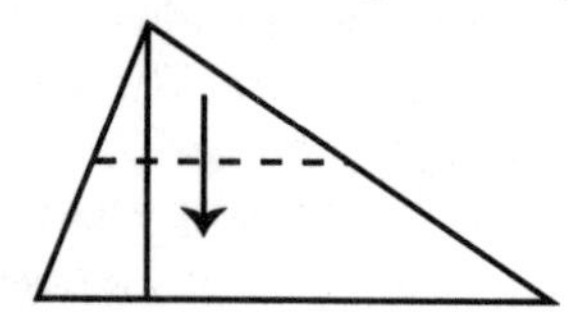
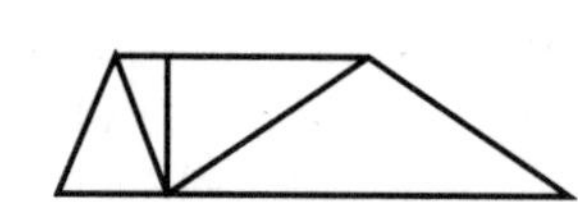

4. Cut along the line just folded and then cut the resulting triangle along the height line.

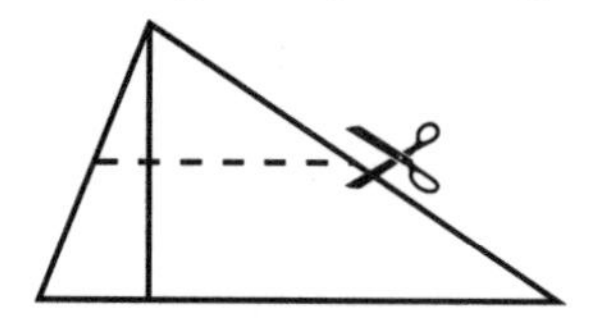

5. Form a rectangle from the three pieces.

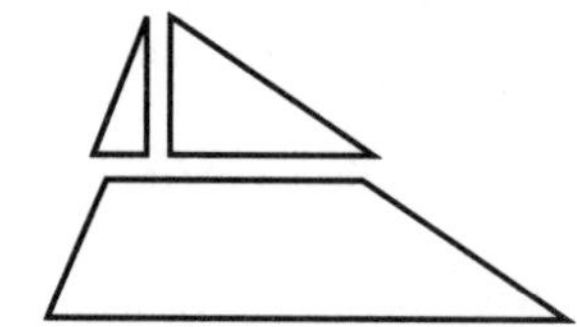
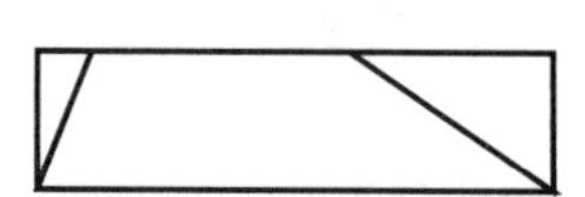

When students have made a rectangle out of one triangle, have them compare it to the remaining triangle and ask them the following questions. *How can you tell the rectangle and the triangle have the same area?* Students should recognize that the rectangle is the triangle which has been cut and rearranged. *How do the bases of the triangle and rectangle compare?* When compared, they are the same. *How does the height of the rectangle compare to the height of the triangle?* Some students may prove the rectangle's height is half the triangle's height by folding the triangle's height in half. When students have observed these relationships ask, *Can you write a formula of the rectangle's area by relating it to the triangle's base and height?* This encourages an alternate equation, Area = base • (height ÷ 2).

Another equivalent formula develops as students follow a different set of directions.

1. Cut out two congruent triangles and place one in a position where the vertex is directly above the base. Again, one triangle will be used for making comparisons.

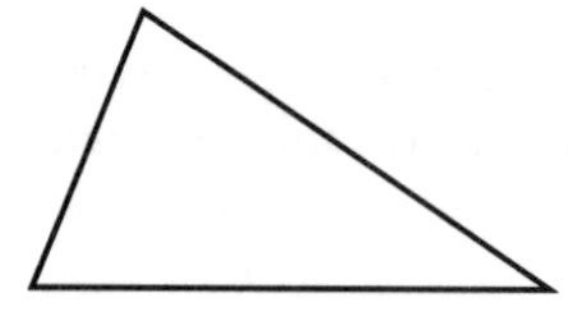
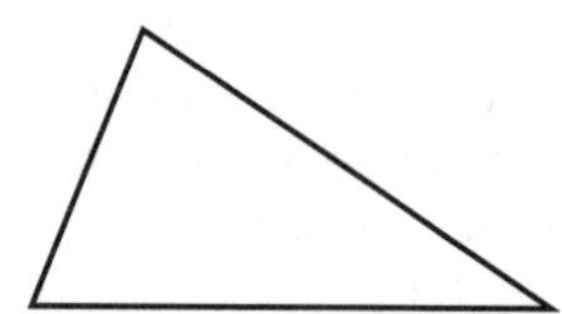

2. Fold the base on top of itself so the fold will be perpendicular to the base. Move the base across itself until the fold goes through the vertex. Crease to form the height line.

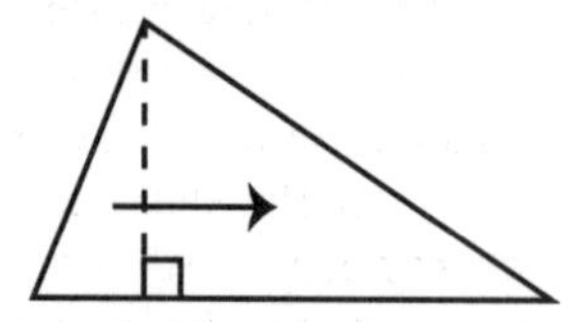

3. Fold each vertex on the base into the height line keeping the base line aligned on top of itself. The

folds will form the perpendicular bisectors of each side of the base.

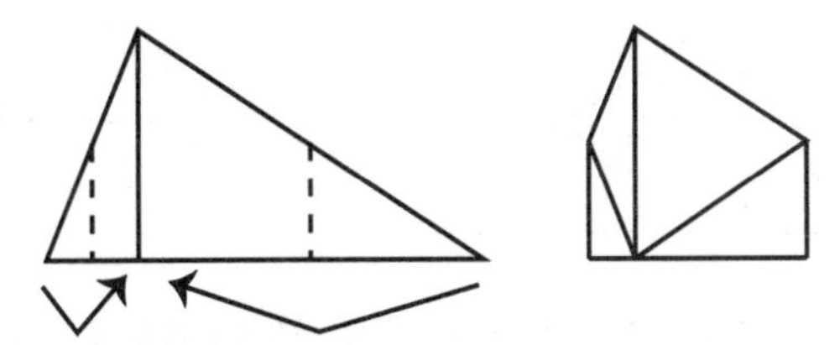

4. Cut along the perpendicular bisectors just folded and use the three resulting pieces to form a rectangle.

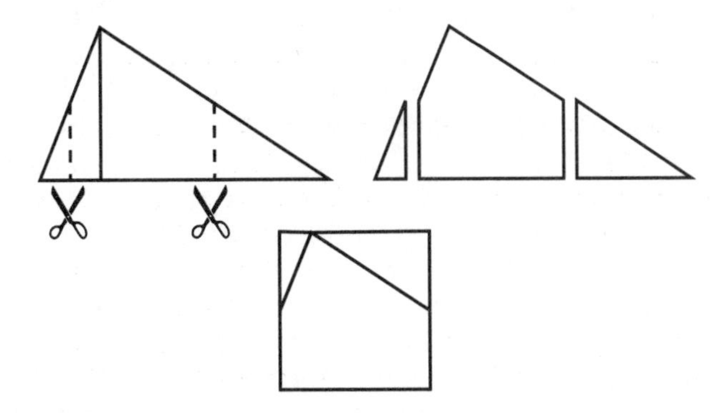

When students have made a rectangle out of one triangle, ask them the following set of questions. *How can you tell the rectangle and the triangle have the same area?* Students should recognize that the rectangle is the triangle, cut and rearranged. *How do the height of the triangle and rectangle compare?* When compared, they are the same. *How does the base of the rectangle compare in length to the base of the triangle?* Some students may prove the rectangle's base is half the triangle's base by folding the triangle's base in half. When students have observed these relationships ask them, *Can you write a formula of the rectangle's area by relating it to the triangle's base and height?* This encourages an alternate equation, Area = (base ÷ 2) • height.

If the class has developed all three equations, it is worthwhile to have a discussion on the equivalency of the formulas.

$$\text{Area} = (\text{base} \bullet \text{height}) \div 2 = \text{base} \bullet (\text{height} \div 2) = (\text{base} \div 2) \bullet \text{height}$$

Encourage students to consider the formulas' similarities and differences. Have them consider ways to prove they are equivalent. They might try substituting the same value base and height into each equation. They might discuss the associative and commutative properties. Also have the students consider when one form might be more appropriate to use than another.

Trapezoids

Students have the least experience with trapezoids. Some exploration about the relationship of trapezoids to other quadrilaterals is a good starting point. Have the students cut rectangles and parallelograms from adding machine tape and consider these two questions. *How many different places can you cut a rectangle one time so the resulting pieces make a trapezoid? How many different places can you cut a parallelogram so the two resulting pieces form a trapezoid?* As students generalize their observations, they should recognize that in both cases the solutions are infinite. For both figures, they can make a cut between opposite sides and flip one piece. For rectangles, the cut must be made so it is not perpendicular to two opposite sides, while the parallelogram requires that the cut be made perpendicular to the sides.

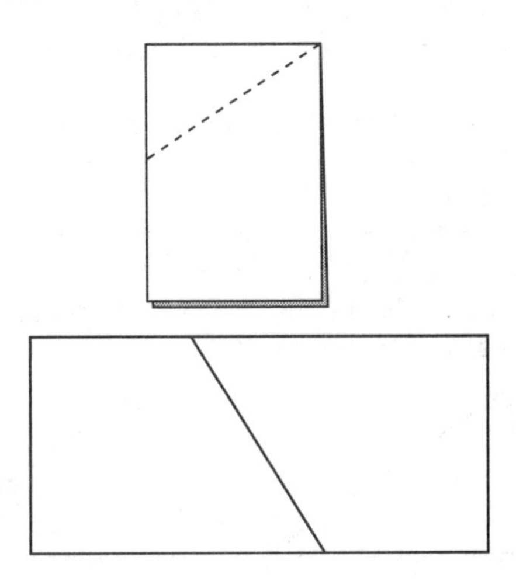

To help students discover a formula for the area of a trapezoid, have them cut two congruent trapezoids by folding a strip of adding machine tape onto itself with the edges aligned.

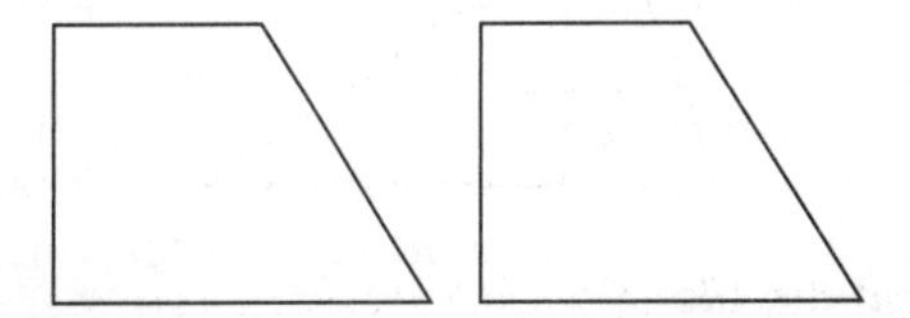

Pose the following question to the students, *Will any two congruent trapezoids form a parallelogram?* They will discover that one of the trapezoids can always be rotated and slid to form a parallelogram. The base of the parallelogram is the length of the top (b_1) and the bottom (b_2). The area of the trapezoid is half the area of the parallelogram so the formula could be written:

$$\text{Area} = \frac{(b_1+b_2) \bullet h}{2}$$

An equivalent formula is developed by having students fold and cut trapezoids as suggested in the following question, *If you cut a trapezoid in half, parallel to the two parallel edges, will you always be able to use the two pieces to form a parallelogram?* As the top part is rotated, the students will discover the two parts will always form a parallelogram that is equal in area but half the height of the original trapezoid.

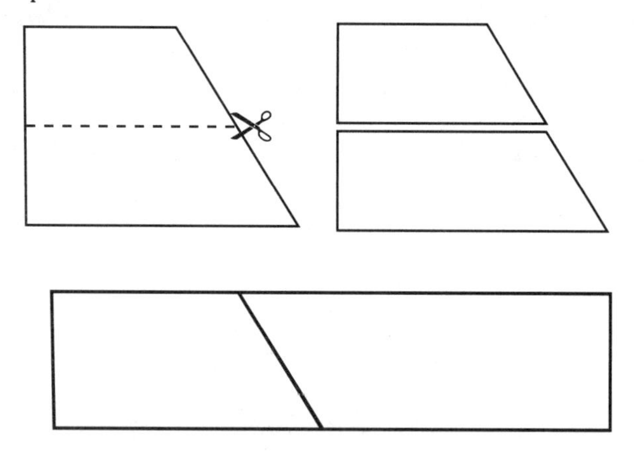

The resulting formula is: Area = $(b_1 + b_2) \bullet \frac{1}{2}h$. Again it would be of value to discuss the equivalency of the formulas and when each might be useful.

Circles

Folding and cutting paper can be used to develop a formula for the area of a circle. The paper used can be inexpensive paper plates, cupcake liners, or round coffee filters. Students must also understand the relationship of the diameter of a circle and its circumference. To be successful, students must be familiar with the formula: Circumference = $\pi \bullet$ (2 $\bullet$ radius) = $\pi \bullet$ diameter. They should already be familiar with the formula for a parallelogram.

To help in the discussion, have the students trace around the outside of the circles (the plates, liners, or coffee filters) with a colored marker. Have the students fold the circles in half and then in half again to make four quarter pie-shaped sections. Have them use a different colored marker to trace one of the radii. Have the students fold each quarter section in half to get eighth sections.

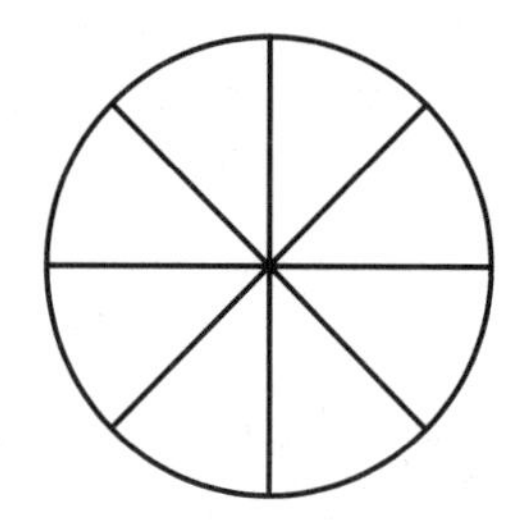

Have the students cut the sections apart and ask, *How can you make a parallelogram out of the eight sections of a circle?*

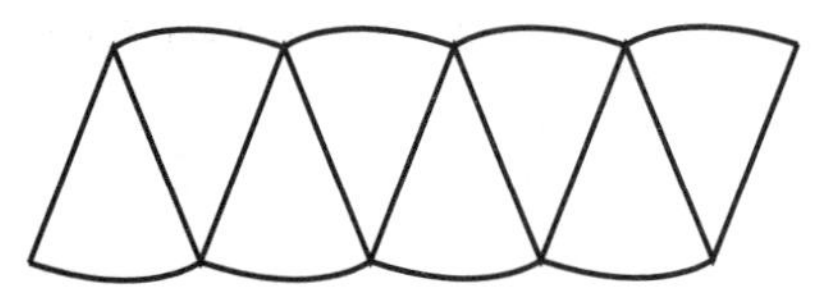

When students have made the parallelogram, discuss the following series of questions:

- *How do you know the circle and the parallelogram have the same area?* [made from the same parts, can be assembled into either]
- *What is the height of the parallelogram relative the circle's dimensions?* [one radius high]
- *How long is the base of the parallelogram relative to the circle's dimensions?* [half of the circumference, $(2\pi r) \div 2 = \pi r$]
- *How could you write a formula to find the area of the parallelogram using the dimensions of the circle?* [Area = base • height = $\pi r \bullet r = \pi r^2$]

Cutting and folding paper provides some unique insights into the relationships between geometric figures. It allows students to develop formulas that have meaning to them and can be remembered visually. It is one medium to use with a variety of others to develop geometric understanding and encourage algebraic thinking.

Topic
Geometric formulas

Key Question
How many different-sized polygons can you make using all five of the puzzle pieces?

Learning Goals
Students will:
- be able to transform one polygon to another using translation, rotation, and reflection of dissected pieces;
- transform parallelograms, trapezoids, and triangles into a rectangle of equal area; and
- develop formulas based on the relationship between the polygon and its equivalent rectangle.

Guiding Documents
Project 2061 Benchmarks
- *Calculate the circumferences and areas of rectangles, triangles, and circles, and the volumes of rectangular solids.*
- *Usually there is no one right way to solve a mathematical problem; different methods have different advantages and disadvantages.*

*NCTM Standards 2000**
- *Describe sizes, positions, and orientations of shapes under informal transformations such as flips, turns, slides, and scaling*
- *Understand relationships among the angles, side lengths, perimeters, areas, and volumes of similar objects*
- *Use geometric models to represent and explain numerical and algebraic relationships*
- *Develop and use formulas to determine the circumference of circles and the area of triangles, parallelograms, trapezoids, and circles*

Math
Geometry
measurement
area
polygon formulas

Integrated Processes
Observing
Comparing and contrasting
Generalizing

Materials
Scissors

Background Information
The fundamental formula of area is average length times average width. Rectangular regions most clearly demonstrate this with constant lengths and width (Area = length • width). All area formulas are modifications on this basic theme.

This puzzle is based on pieces that form a two unit by six unit rectangle or a three unit by four unit rectangle.

Rectangle Key

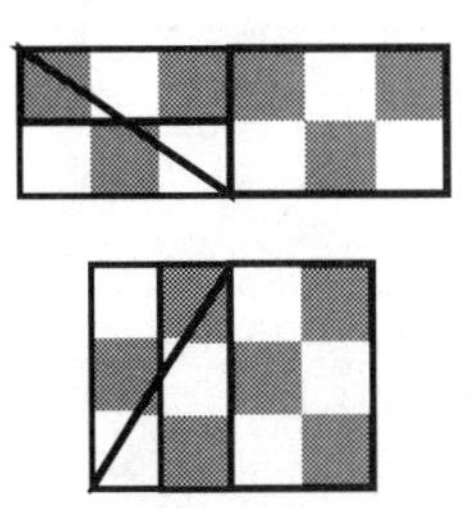

The five puzzle pieces can be translated (slid), rotated (spun), or reflected (flipped) and rearranged to form two parallelograms, two trapezoids, and a triangle. These shapes all have different dimensions but the same area as the rectangles—12 square units. As students work with the puzzle, they will begin to recognize that each polygon can be transformed into one of the rectangles simply by moving pieces. As the polygon is transformed into the rectangle, students can compare the dimensions of the polygon to its corresponding rectangle. The basic length times width formula of the rectangle is transformed to relate one of the rectangle's two dimensions to one of the polygon's dimensions. Students begin to develop the understanding that all area mesurement is two dimensional, a function of the length and width of the polygon.

The parallelograms are transformed by cutting off a triangle on one end and translating it to the other end to form the rectangle. The base and height of the parallelogram and its rectangle are equal. The fundamental formula (Area = base • height) remains unchanged except for the change in terms to clarify the dimensions of the parallelogram.

Parallelogram Key

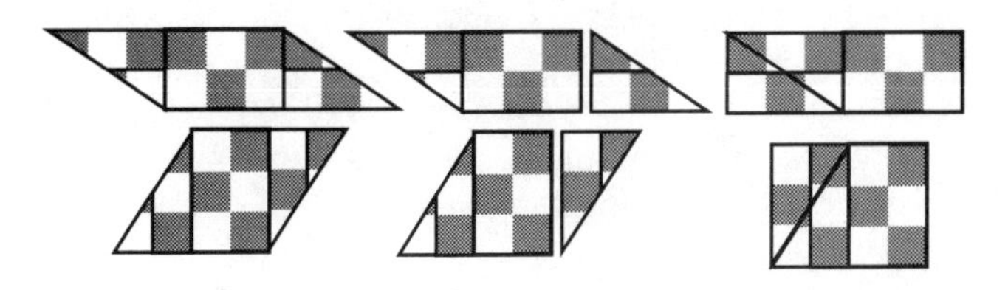

The trapezoids are transformed by cutting off a triangle on one side and rotating it into place on the opposite end. What is taken from the longer base is given to the shorter base. The lengths of the bases have been averaged. Students often describe this average as "being in the middle of the two lengths." The height of the trapezoid and its rectangle are the same. The resulting formula shows this averaging of the bases and the equal heights. (Area = $(b_1 + b_2 \div 2) \bullet$ height)

Trapezoid Key

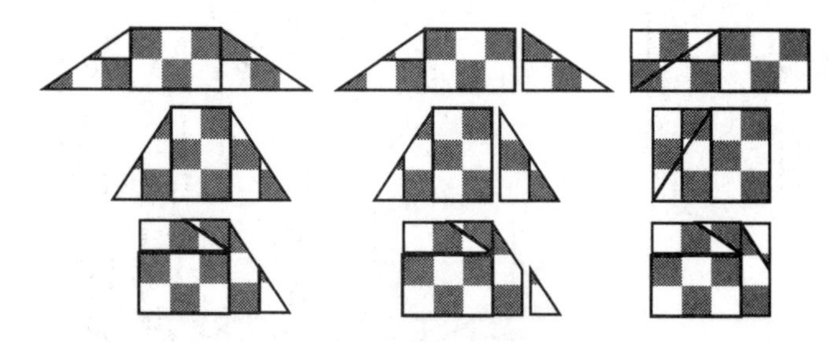

The triangle can be transformed into a rectangle in two ways. First, a triangle can be cut off one side that is half the length of the original triangle and then rotated onto the top of the remaining portion. The second way is to cut a triangle off the top that is half the height of the original triangle. This piece is then rotated onto the side. Both methods leave one dimension of the rectangle and triangle the same, while the other dimension of the triangle is half the length of the corresponding dimension of the rectangle. The two formulas show this similarity.

(Area = $\frac{\text{length}}{2} \bullet$ height) (Area = $\frac{\text{height}}{2} \bullet$ length)

Triangle Key

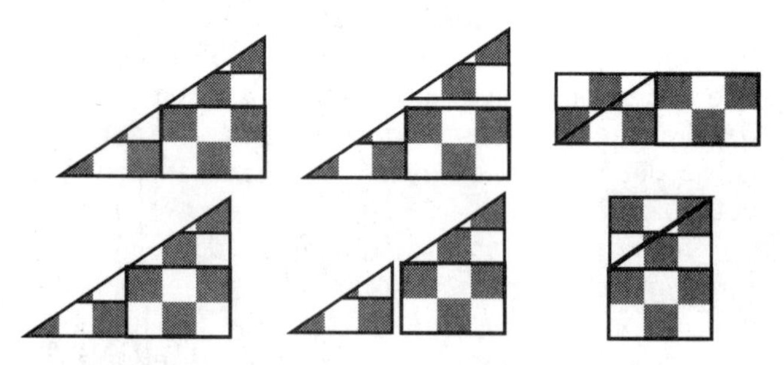

Management

1. Students need time to explore and find a variety of solutions. Have them construct the puzzle at the beginning of a week and then develop the formulas at the end of the week when everyone has been successful at finding most of the possible polygons.
2. This works well as an individual puzzle with collaboration taking place as students share hints of how to find possible solutions.

Procedure

1. Distribute puzzles and direct the students to cut the pieces out with scissors.
2. Allow time for them to explore, discover, and record polygons that can be made with all five pieces. Emphasize to students that they must use all five pieces and that the pieces may be translated, reflected, and rotated.
3. Challenge them to determine the fewest pieces that need to be moved in order to transform each of their parallelograms, trapezoids, and triangles into rectangles. Have them record their solutions.
4. Discuss with the students the similarities in the polygons and their corresponding rectangles. Direct them to determine a formula to describe the relationships found.

Discussion

1. How are all the polygons you found the same? [equal area]
2. How are the dimensions of each type of polygon similar to and different from the rectangles that you made from them? (Refer to *Background Information.)*
3. How could you describe the dimension of the rectangle relative to the dimension of the polygon from which is was made? (Refer to *Background Information.)*
4. How would you write a formula to find the area of each type of polygon? (Refer to *Background Information.)*

Extension

Cut out parallelograms, trapezoids, and triangles. Have students measure them and calculate the areas. Have the students cut them and transform them into rectangles and check to see if the rectangles have the same areas as the polygons from which they were made.

Cut out the five pieces of the puzzle. See how many different polygons can be constructed using all five pieces. Pieces may be turned upside down to make a solution.

Cut out the five pieces of the puzzle. See how many different polygons can be constructed using all five pieces. Pieces may be turned upside down to make a solution.

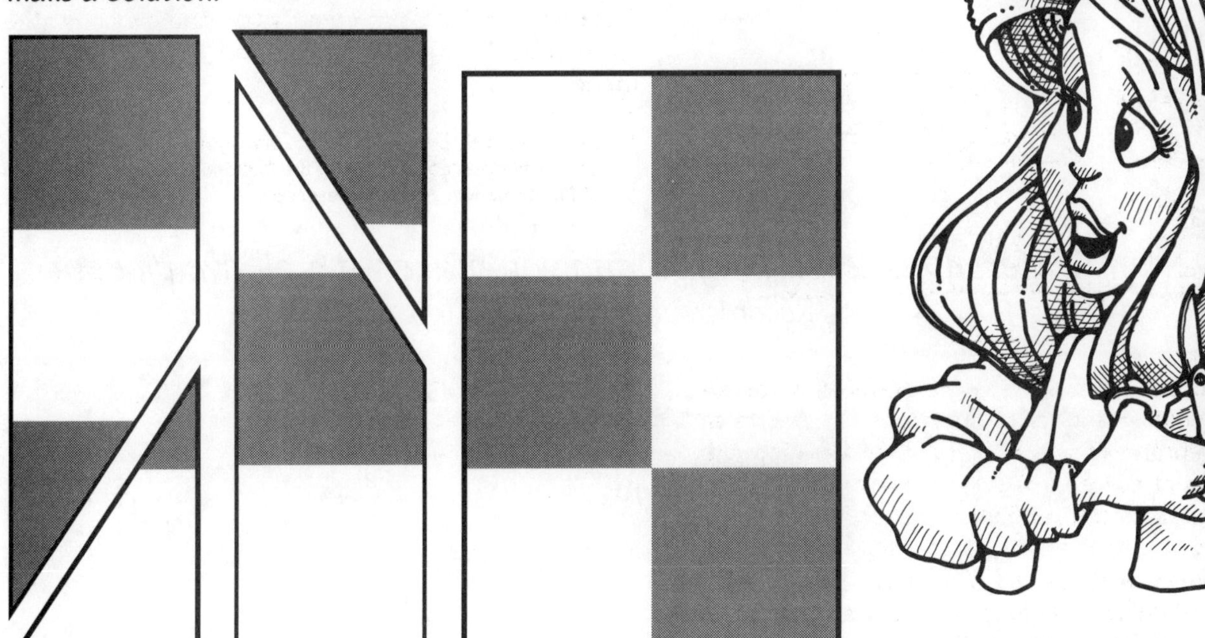

How many different-sized rectangles can you make with all five pieces?

Record your solutions.

How many different-sized parallelograms can you make with all five pieces?

Record your solutions.

How many different-sized trapezoids can you make with all five pieces?

Record your solutions.

How many different-sized triangles can you make with all five pieces?

Record your solutions.

What are the fewest pieces you need to move to change a parallelogram into a rectangle?

Record each of the parallelograms and the rectangle into which it is transformed.

Record the dimensions of the rectangle in relation to the base(b) and height(h) of the parallelogram.

What are the fewest pieces you need to move to change a trapezoid into a rectangle?

Record each of the trapezoids and the rectangle into which it is transformed.

Record the dimensions of the rectangle in relation to the bases(b) and height(h) of the trapezoid.

What are the fewest pieces you need to move to change a triangle into a rectangle?

Record the triangle and two rectangles into which it can be transformed.

Record the dimensions of the rectangles in relation to the base(b) and height(h) of the triangle.

Topic
Algebra, Geometry—Pythagorean Theorem

Key Questions
1. If you know the lengths of the three sides of a triangle, how can you determine if the triangle is equilateral, isosceles, or scalene?
2. If you know the lengths of the three sides of a triangle, how can you determine if the triangle is acute, right, or obtuse?

Learning Goals
Students will:
- learn to sort triangles by angles and sides, and
- recognize the equality and inequalities between the sum of the areas of the squares on the short legs of a triangle compared to the area of the square on the longest side

Guiding Documents
Project 2061 Benchmarks
- *Some shapes have special properties: Triangular shapes tend to make structures rigid, and round shapes give the least possible boundary for a given amount of interior area. Shapes can match exactly or have the same shape in different sizes.*
- *Mathematicians often represent things with abstract ideas, such as numbers or perfectly straight lines, and then work with those ideas alone. The "things" from which they abstract can be ideas themselves (for example, a proposition about "all equal-sided triangles" or "all odd numbers").*
- *Graphs can show a variety of possible relationships between two variables. As one variable increases uniformly, the other may do one of the following: always keep the same proportion to the first, increase or decrease steadily, increase or decrease faster and faster, get closer and closer to some limiting value, reach some intermediate maximum or minimum, alternately increase and decrease indefinitely, increase and decrease in steps, or do something different from any of these.*

*NCTM Standards 2000**
- *Understand relationships among the angles, side lengths, perimeters, areas, and volumes of similar objects*
- *Use geometric models to represent and explain numerical and algebraic relationships*
- *Use symbolic algebra to represent situations and to solve problems, especially those that involve linear relationships*
- *Use graphs to analyze the nature of changes in quantities in linear relationships*
- *Create and critique inductive and deductive arguments concerning geometric ideas and relationships, such as congruence, similarity, and the Pythagorean relationship*

Math
Geometry
  categories of triangles
  Pythagorean Theorem
Algebra
  graphing
  equalities/inequalities

Integrated Processes
Observing
Comparing and contrasting
Generalizing

Materials
Option One:
  card stock
  scissors
Option Two:
  AIMS Algebra Tiles
  2 cm squares

Background Information
Finding all the possible triangular regions that can be surrounded from a limited variety of discrete size squares can provide a rich understanding of triangles.

Initially students will discover that not all combinations of three squares work. In order to form a triangle, the combined edges of two shorter squares must be greater than the length of the edge of the largest square. With the four different size squares of 2, 3, 4, and 5 centimeters, 20 different combinations can be made. Three of these, however, will not form triangles (2,2,4; 2,2,5; 2,3,5).

Constructing 17 different triangles provides an excellent opportunity to reinforce the understanding of types of triangles. When sorting by sides, students will find four equilateral, three scalene, and ten isosceles triangles. When sorting by angles, students will be able to find one right, four obtuse, and 12 acute triangles.

Having students consider the sum of the areas of squares made on the two shorter legs of a triangle ($a^2 + b^2$) compared to the area of the square formed on the longest leg (c^2) develops a full concept of the Pythagorean theorem ($a^2 + b^2 = c^2$). Placing these measurements alongside the types of triangles by angles quickly shows the following:

when $a^2 + b^2 > c^2$, it is an acute triangle;
when $a^2 + b^2 = c^2$, it is a right triangle;
when $a^2 + b^2 < c^2$, it is an obtuse triangle.

Discovering these equalities and inequalities helps students understand that the Pythagorean theorem is true only for right triangles. It also provides a way to interpret the broader relationship of ($a^2 + b^2$) to (c^2).

By graphing the measurements, the relationship is reinforced. As students plot each triangle's point with ($a^2 + b^2$) being the vertical axis and (c^2) being the horizontal axis, they will discover that only the right triangle falls on the line with a slope of one. The acute triangles all plot above the one line and the obtuse triangles below the line. This accentuates the special relationship of the right triangle while providing a very pertinent use of graphing inequalities.

Management

1. Before the lesson, determine what materials will be used for squares. *Option One:* The blackline master may be copied onto card stock and students may cut out the squares when they are needed. *Option Two:* AIMS Algebra Tiles, which provide a more kinesthetic experience for students, may be used; however, the teacher will need to cut 2 cm square pieces before beginning the lesson.
2. Students should be familiar with the types of triangles by sides (equilateral, isosceles, scalene) and angle (acute, right, obtuse) before beginning this investigation. If they are not, some instruction should be done to familiarize them with all the types.
3. Emphasize to the students that they need to be careful when forming the triangles. The corners of the squares must touch each other to form a vertex. Several of the acute and obtuse triangles will be mistakenly classified as right triangles if they are not constructed accurately. A corner from a piece a paper will help students check right angles.
4. Partners work very well for this activity. One student can construct while the second records.

Procedure

1. Distribute the materials and allow students time to cut out the squares if necessary. If the materials are not labeled for length and area, have the students examine and compare the pieces to recognize the sizes are 2, 3, 4, and 5 centimeters.
2. Using the squares have students determine how many different-sized triangles can be made from these four sizes of squares.

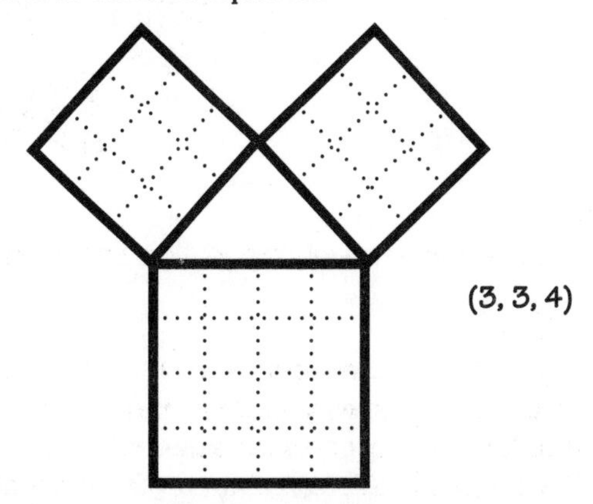

Although many students will begin by randomly developing combinations, encourage them to develop an ordered listing to assure they have found all the possible combinations. They should discover that three combinations of squares do not form triangles (2,2,4; 2,2,5; 2,3,5).

3. When students have confirmed that a combination of squares forms a triangle, direct them to record the lengths on the chart.
4. Encourage the students to carefully form the triangles so the corners of the squares touch exactly. Have them determine and record the type of triangle that is formed. (A corner of a paper can be used to determine right angles.)
5. Optional: Have students cut squares of paper and glue them to make permanent records of all possible triangles.
6. Direct the students to calculate the area of the square on each edge of the triangle. Have them record the sum of the areas of the squares of the two shorter legs in one column of the chart and the area of the square on the longest leg in another column of the chart.

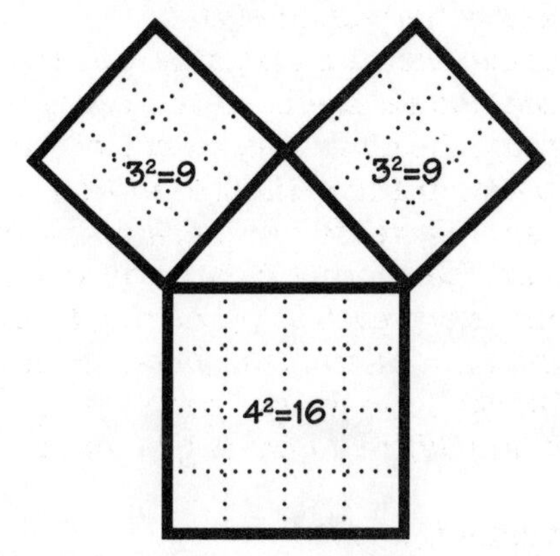

7. Tell them to choose the equality/inequality sign that makes the number sentence true.
8. Hold a discussion in which students share observations they make about the relationships of number sentences and types of triangles.

9. Have them make a graph of the data by plotting the areas of the squares on the sides of a triangle.
10. Have them observe and record the relationship of the position where each triangle is represented on the graph and the type of triangle it is. Direct them to record the algebraic equation that represents the line that is drawn on the graph [$a^2 + b^2 = c^2$ or = 1] and inequalities describing the regions above [$a^2 + b^2 > c^2$ or > 1] and below [$a^2 + b^2 < c^2$ or < 1] the line.

Discussion

1. How can you tell if a combination of squares will make a triangle before assembling it? [To make a triangle, the combination of the lengths of the two shorter sides must be greater than the length of the longest side ($a + b > c$).]
2. How can you look at the lengths of the sides of a triangle and determine what type of triangle it is by sides?
3. What relationship do you see in squares on the sides of a triangle and the type of triangle it is by angles? [$a^2 + b^2 > c^2$: acute; $a^2 + b^2 = c^2$: right; $a^2 + b^2 < c^2$: obtuse]
4. What is the relationship of the position of a triangle's point on the graph and the type of triangle it is by angle? [right on the one line, acute above the one line, obtuse below the one line]
5. If you know the lengths of the edges of a triangle, how can you determine what type of a triangle it is by angles? [Find the squares on all three sides. If the sum of the squares on the shortest two sides equals the square on the longest side, it is a right triangle. If the sum is greater, it is an acute triangle. If the sum is less, it is an obtuse triangle.]

Extension

Measure triangles found in the classroom and use the algebraic sentences to determine their types. Confirm that it is correct.

* Reprinted with permission from *Principles and Standards for School Mathematics,* 2000 by the National Council of Teachers of Mathematics. All rights reserved.

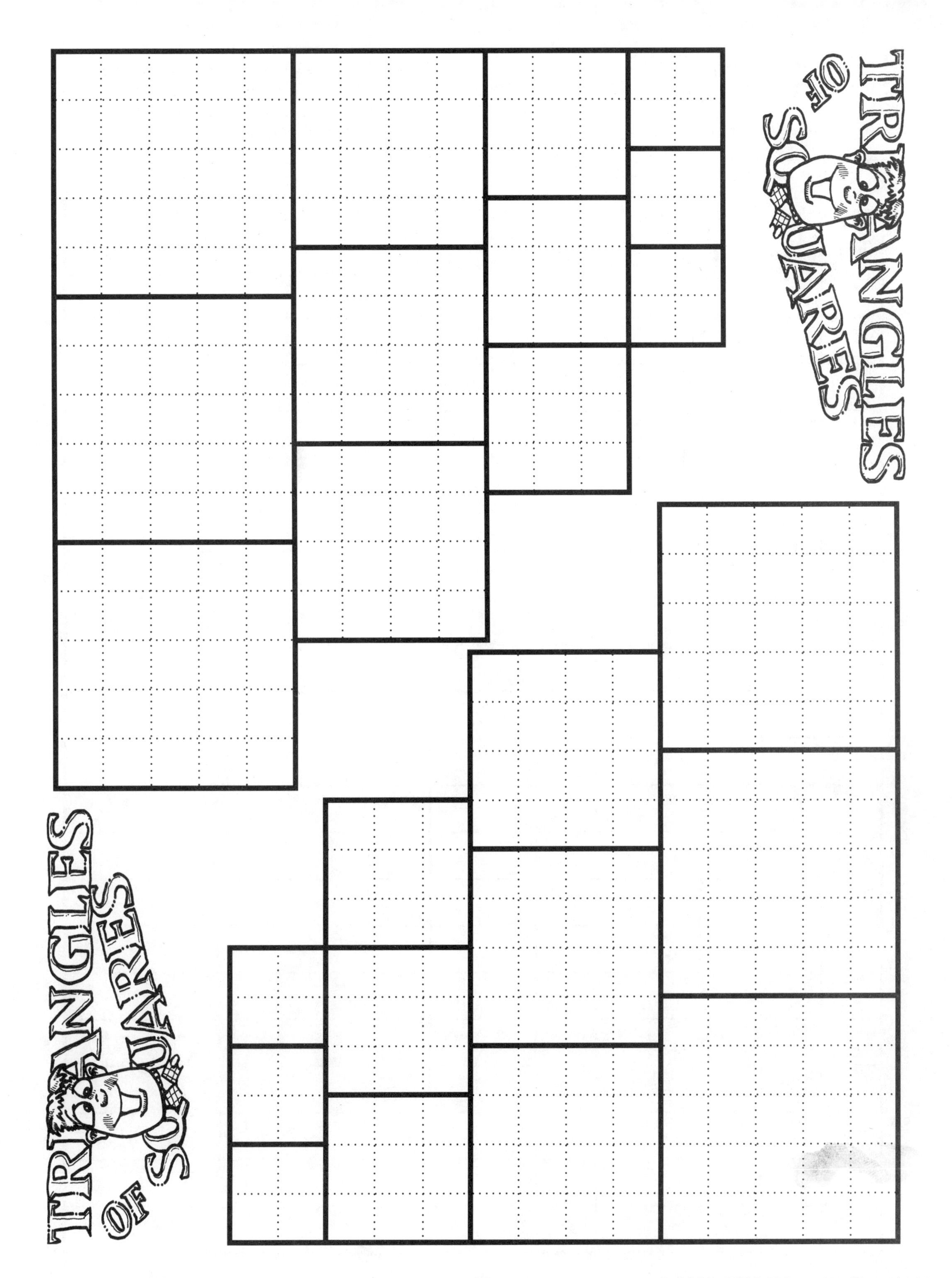
TRIANGLES OF SQUARES
TRIANGLES OF SQUARES

Make all the possible different size triangles you can out of 2,3,4, and 5 cm squares. Record the dimensions and complete the table.

Length of Sides			Squares on Sides			Type of Triangle	
Shortest Side a	Mid-length Side b	Longest Side c	Sum of squares on shorter sides a^2+b^2	< > =	Area of square on longest side c^2	By Angles Acute Right Obtuse	By Sides Equilateral Isosceles Scalene

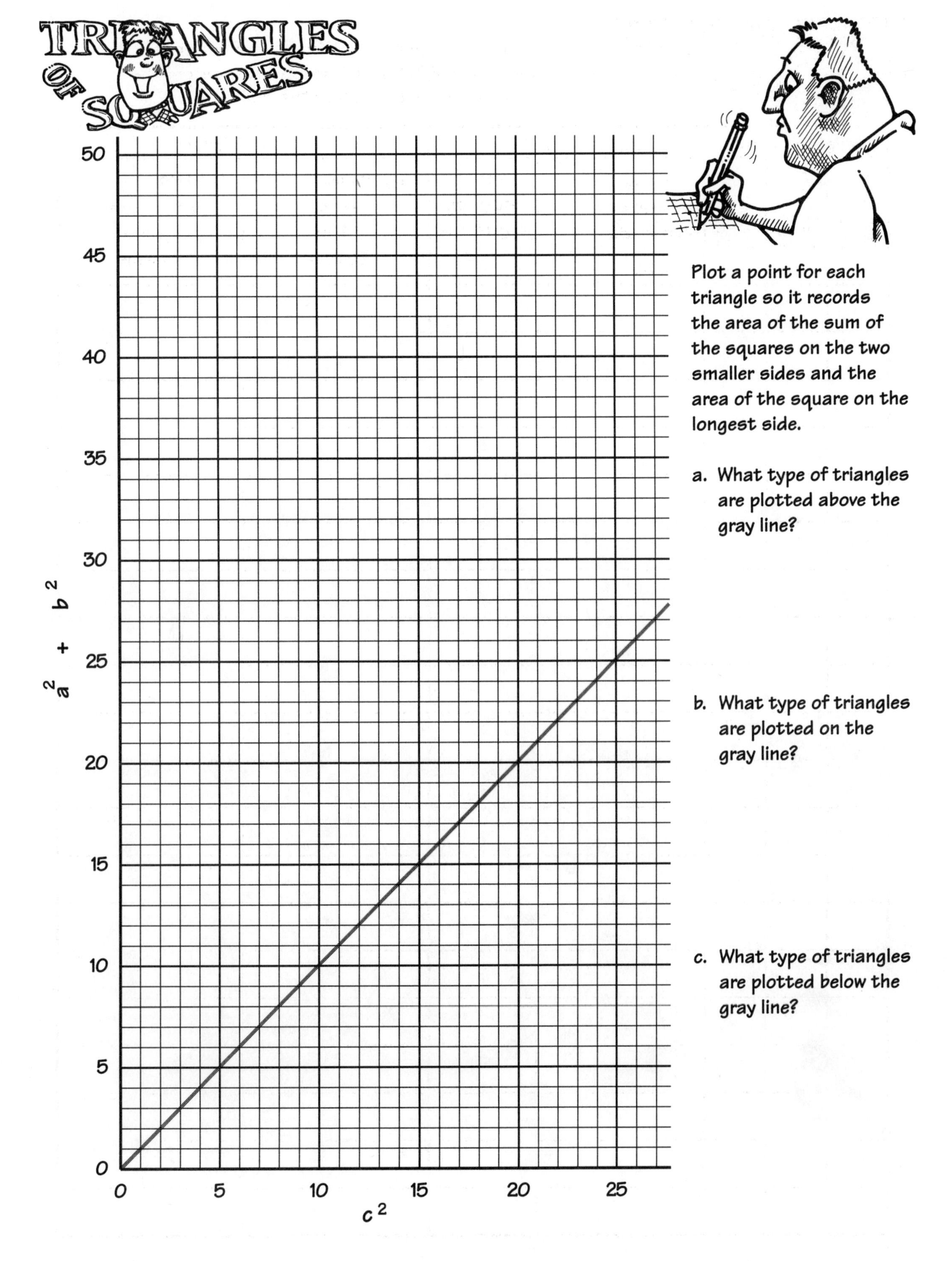
TRIANGLES OF SQUARES
Plot a point for each triangle so it records the area of the sum of the squares on the two smaller sides and the area of the square on the longest side.
a. What type of triangles are plotted above the gray line?
b. What type of triangles are plotted on the gray line?
c. What type of triangles are plotted below the gray line?
$a^2 + b^2$
c^2
0
5
10
15
20
25
30
35
40
45
50

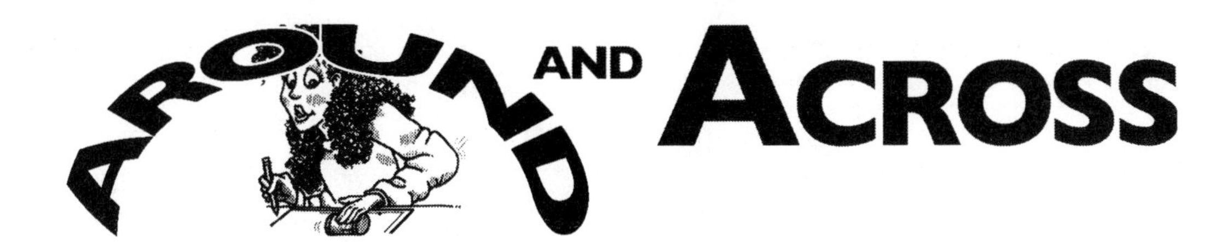

Topic
Circumference to Diameter Relationship (pi)

Key Question
How does the distance around the circle (circumference) compare to the distance across a circle (diameter)?

Learning Goals
Students will:
- generate a graph by using the diameters and circumferences of round objects,
- study the scatter plot to discover the proportional relationship,
- use numeric data from the graph to approximate the value of pi, and
- determine a formula for the circumference of a circle.

Guiding Documents
Project 2061 Benchmarks
- *The graphic display of numbers may help to show patterns such as trends, varying rates of change, gaps, or clusters. Such patterns sometimes can be used to make predictions about the phenomena being graphed.*
- *Graphs can show a variety of possible relationships between two variables. As one variable increases uniformly, the other may do one of the following: always keep the same proportion to the first, increase or decrease steadily, increase or decrease faster and faster, get closer and closer to some limiting value, reach some intermediate maximum or minimum, alternately increase and decrease indefinitely, increase and decrease in steps, or do something different from any of these.*

NRC Standards
- *Use mathematics in all aspects of scientific inquiry.*
- *Use appropriate tools and techniques to gather, analyze, and interpret data.*
- *Develop descriptions, explanations, predictions, and models using evidence.*

*NCTM Standards 2000**
- *Use representations to model and interpret physical, social, and mathematical phenomena*
- *Represent, analyze, and generalize a variety of patterns with tables, graphs, words, and, when possible, symbolic rules*
- *Model and solve contextualized problems using various representations, such as graphs, tables, and equations*
- *Develop and use formulas to determine the circumference of circles and the area of triangles, parallelograms, trapezoids, and circles and develop strategies to find the area of more complex shapes*

Math
Measurement
Geometry
  pi (π)
  $2 \pi r = C$
Proportional reasoning
Statistics
  scatter plot
Algebra
  slope
  equations

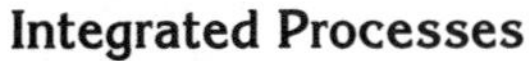

Integrated Processes
Observing
Collecting and organizing data
Interpreting data
Generalizing

Materials
Metric rulers or meter sticks
Variety of round objects (cans, container lids)
chart or butcher paper.

Background Information
Pi (π) is the relationship of the circumference of a circle to its diameter. It is an irrational number. Its approximation is written as 22/7 as a fraction or 3.1415... as its decimal equivalent. The approximation of 3.14 is sufficient in all but the most precise applications.

Students are often introduced to the circumference of circles with the formula $C=2\pi r$. They do not have any experience for making connections to the formula so it tends to be meaningless and quickly forgotten or mixed up with other formulae.

Using round objects to make a concrete graph provides a valuable experience in which students can think about the relationship of circumference and diameter. They can come to see that the diameter and circumference increase proportionately. They can use the data from the graph to verify this graphically, numerically, and symbolically.

Management
1. Students need to be familiar with measuring in centimeters to do this activity.
2. This activity can be completed in a 45-60 minute period.

3. Before the activity, gather a variety of different- sized round objects such as cans, lids, toy car tires, and other items.

Procedure

1. Discuss the *Key Question* and state the *Learning Goals.*
2. Have students construct a graph on a piece of chart paper or butcher paper. Guide them to use a straight edge to draw a horizontal axis just up from the bottom edge of the paper and label it *diameter.* Have them make a vertical axis just in from the left edge of the paper and label it *circumference.* (The intersection of the two axes in the lower left corner of the paper is the origin.)

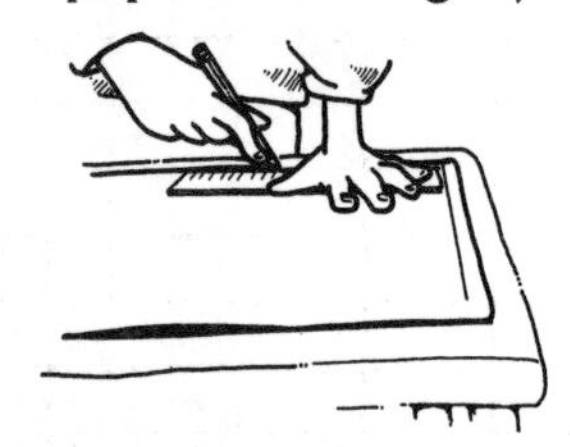

3. Inform the students that for each round object, they must find the point on the horizontal axis that corresponds with that object's diameter. Show them that they can do this by putting the diameter of the round object on the horizontal axis with one end on the origin. Have them mark the other endpoint along the axis to show the distance of the diameter from the origin.

4. Instruct the students to make a small line on the circumference of the round object. Have them place this line on top of the dot they placed on the horizontal axis which indicated the object's diameter. Inform them that the object must be positioned so that it can be rolled up the paper, parallel to the vertical axis.

5. Instruct the students to roll the round object one complete revolution until the mark is back on the paper. Have them mark a point at this position on the paper. (This point represents the relationship between the circumference and diameter of the can.) Caution students to keep the round object rolling parallel to the vertical axis.
6. Have the students graph the points for all of the round objects.

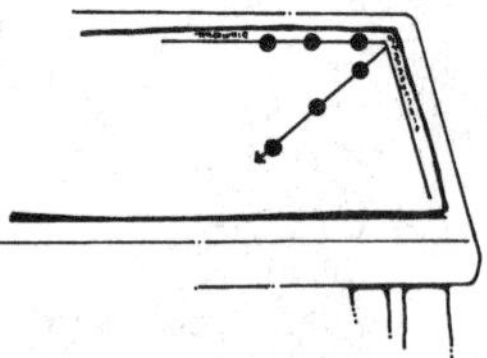

7. Discuss what patterns can be seen in the graph and what information it gives about round objects. [straight line, goes up as it goes right; circumference is always bigger than diameter, bigger diameters result in bigger circumferences]
8. Have the students measure and record the diameters and circumferences of the round objects using the graph.
9. Direct them to calculate and record the ratio of circumference to diameter for each of the round objects, and then average the ratios.
10. Discuss how the ratios compare to each other, what that says about circles, and how it is related to the graph.
11. Have the students write an equation to summarize what they have learned about circles.

Discussion

1. What patterns do you see in the graph? [straight line, goes up as it goes right]
2. What do the patterns tell you about circles? [circumference is always bigger than diameter, bigger diameters result in bigger circumferences, circumference grows proportionally to diameter]
3. What is the average ratio of circumference to diameter for all the objects? [should approach pi, 3.14]
4. What does this average tell you about circles? [how many times bigger the circumference is than the diameter]
5. How is the average related to the graph? [tells how much the lines go up for each step to the right, the slope]
6. How could you determine the circumference of a circle if you knew the diameter? [use the graph by interpolating or extrapolating, multiply the diameter by the average]
7. Write an equation you can use to determine the circumference of a circle if you know the diameter. [π (d) = C]

Extension

Measure the diameters of large circles which have been painted on athletic fields and predict their circumferences. Then measure them to see if they are correct.

How does the circumference of a circle compare to the diameter of a circle?

Round Object	Diameter	Circumference	Ratio $\frac{c}{d}$
		Average	

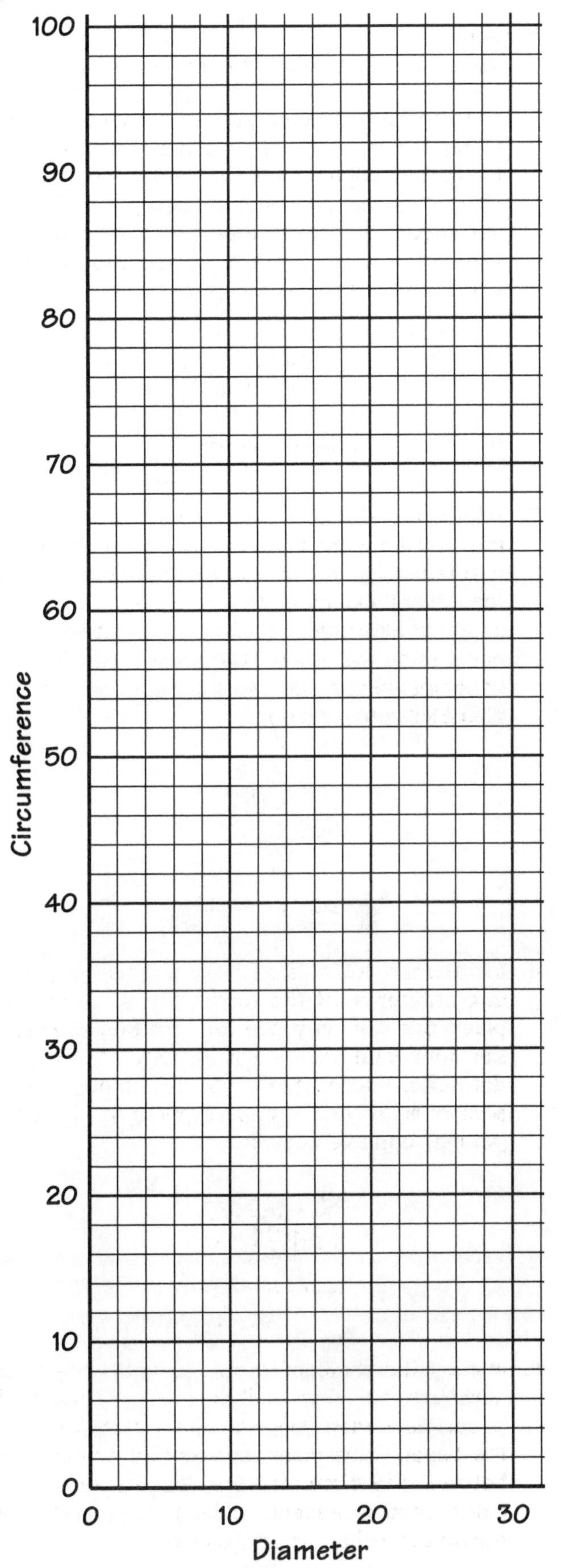

Topic
Areas of Circles

Key Question
How does the area of a circle relate to the length of the circle's radius?

Learning Goals
Students will:
- measure by counting the lengths of the radius of circles, the area of the circles, and area of the radius squares of the circles;
- graph the data to discover the area/radius relationship; and
- use the numeric data to determine a formula.

Guiding Documents
Project 2061 Benchmarks
- *The graphic display of numbers may help to show patterns such as trends, varying rates of change, gaps, or clusters. Such patterns sometimes can be used to make predictions about the phenomena being graphed.*
- *Graphs can show a variety of possible relationships between two variables. As one variable increases uniformly, the other may do one of the following: always keep the same proportion to the first, increase or decrease steadily, increase or decrease faster and faster, get closer and closer to some limiting value, reach some intermediate maximum or minimum, alternately increase and decrease indefinitely, increase and decrease in steps, or do something different from any of these.*

NRC Standards
- *Use mathematics in all aspects of scientific inquiry.*
- *Use appropriate tools and techniques to gather, analyze, and interpret data.*
- *Develop descriptions, explanations, predictions, and models using evidence.*

*NCTM Standards 2000**
- *Use representations to model and interpret physical, social, and mathematical phenomena*
- *Represent, analyze, and generalize a variety of patterns with tables, graphs, words, and, when possible, symbolic rules*
- *Model and solve contextualized problems using various representations, such as graphs, tables, and equations*
- *Develop and use formulas to determine the circumference of circles and the area of triangles, parallelograms, trapezoids, and circles and develop strategies to find the area of more complex shapes*

Math
Measurement
Geometry
- pi (π)
- $\pi r^2 = C$

Proportional Reasoning
Statistics
- scatter plot

Algebra
- slope
- equations

Integrated Processes
Observing
Collecting and organizing data
Interpreting data
Generalizing

Materials
Grid paper
Scissors
Straight edge (rulers)
Variety of round objects (cans, container lids)

Background Information

Pi (π) is the relationship of the circumference of a circle to its diameter. It is an irrational number. Its approximation is written as 22/7 as a fraction or 3.14159... as its decimal equivalent. The approximation of 3.14 works very well in all but the most precise applications.

Pi is a part of formulae of many measurements dealing with circles. It shows up in the formula for the area of the interior region of a circle ($\pi r^2 = A$). It is also a part of the formula for the volume of a sphere ($V = (4\pi r^3)/3$).

Students are often introduced to the formula for the area of a circle, $\pi r^2 = C$, but because they do not have any experience to connect with the formula, it tends to be meaningless and is quickly forgotten or mixed up with other formulae.

By physically counting the area of the radius square, students recognize the meaning of area, and see a visual model of the meaning of a radius squared. By making a graph of the data, they discover the consistent relationship of the circle's area and its radius square's area. When students calculate this relationship as a ratio, they are quick to recognize it as very close to pi. From their experience, they can generate the formula for finding the area of circle.

By generating a graph comparing the circle's area to its radius, and a graph comparing the circle's area to the radius square's area, students are given an opportunity to interpret several graphs. They will recognize the significance of linear patterns. By comparing the graphs and measurements, students can recognize that areas grow proportionally to corresponding areas, while area grows exponentially compared to corresponding linear measures.

Management

1. This activity will take two 45-60 minute periods: one to draw, cut, and measure and the second to graph the data and interpret.
2. Before the activity, gather a variety of different-sized round objects, such as cans, lids, toy car tires, etc.

Procedure

1. Discuss the *Key Question*.
2. Distribute the round objects and blank paper. Have students trace around the object onto paper and cut out the resulting circle.

3. Tell the students to fold the circle into fourths by bringing the opposites sides together and creasing to form a diameter. Then have them bring the opposite ends of the diameter together to form a quarter circle having two radii and an arc.

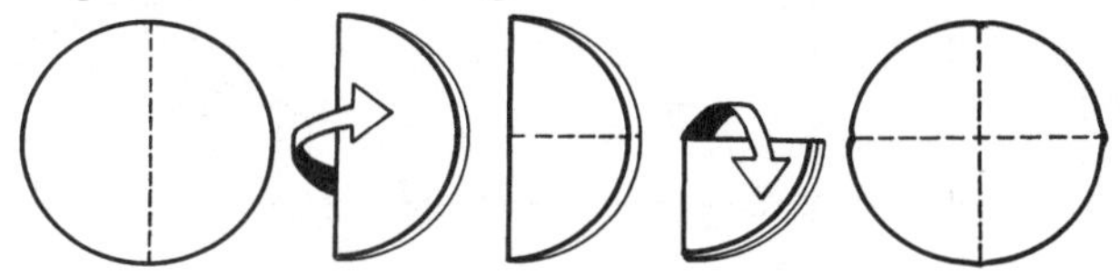

4. Direct the students to trace the circle onto grid paper. To help in counting, have them align the radii within the grid lines so the center is at an intersection of lines.
5. Tell the students to count and record the area of the circle on the grid paper. (This is an estimation process so it may be necessary to discuss the concept of placing several partial squares together to make a whole square.) Hint: The students may find it quicker to count one-fourth of the circle and then multiply by four to get the total.
6. Ask the students to trace the radii and use a straight edge to draw in the remaining two sides of the radius square.
7. Have them count and record the area of the square.

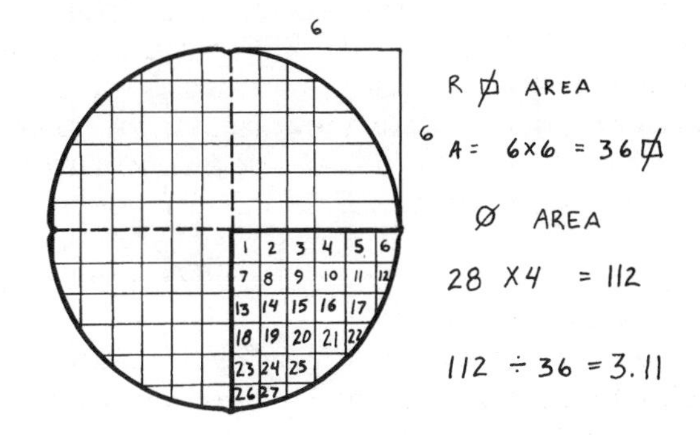

8. Inform the students that they will be making two graphs from the data: One graph compares the circle's area to its radius; the other graph compares the area of the circle to the area of its radius square.
9. Have the students discuss the differences in the graphs.
10. Direct them to calculate and record the decimal ratios of the circles' areas to their radii, and the circles' areas to the area of their radius squares.
11. Have the students discuss what patterns they see in the ratios and how they relate to their corresponding graphs.
12. Invite the students to discuss how they could go about determining the area of a circle if they knew its radius and then write a formula to describe their procedure.

Discussion

1. What are the similarities and differences in the patterns in the two graphs? [Both go up as you go to the right. One is a straight line while the other is a curve.]
2. What patterns do you see in the ratio comparing the area of a circle to its radius? [The ratio gets bigger as the circles get bigger.]
3. How does this number pattern show up on the graph? [The curve gets steeper as the ratio gets larger.]
4. What patterns do you see in the ratio comparing the area of a circle to the area of a square with the circle's radius on each side? [The ratios are very close to the same no matter the size of the circle, all are about 3.]
5. How does this number pattern show up on the graph? [A straight line is formed with a slope of about 3.]
6. What does the ratio tell you about the area of the circle compared to the area of the square whose sides are the length of the circle's radius? [The circle's area is a little more than 3 of the square's area.]
7. What number in math has a value of about 3 that has to do with circles? [pi, 3.14]
8. Write a formula that tells you the area of a circle if you know the length of its radius. [$\pi r^2 = A$]

Extension

Have students draw circles on grid paper and draw squares around the circles. Have them measure and determine the relationship of the area of a circle to the area of its diameter square. [$A = 0.785 d^2$]

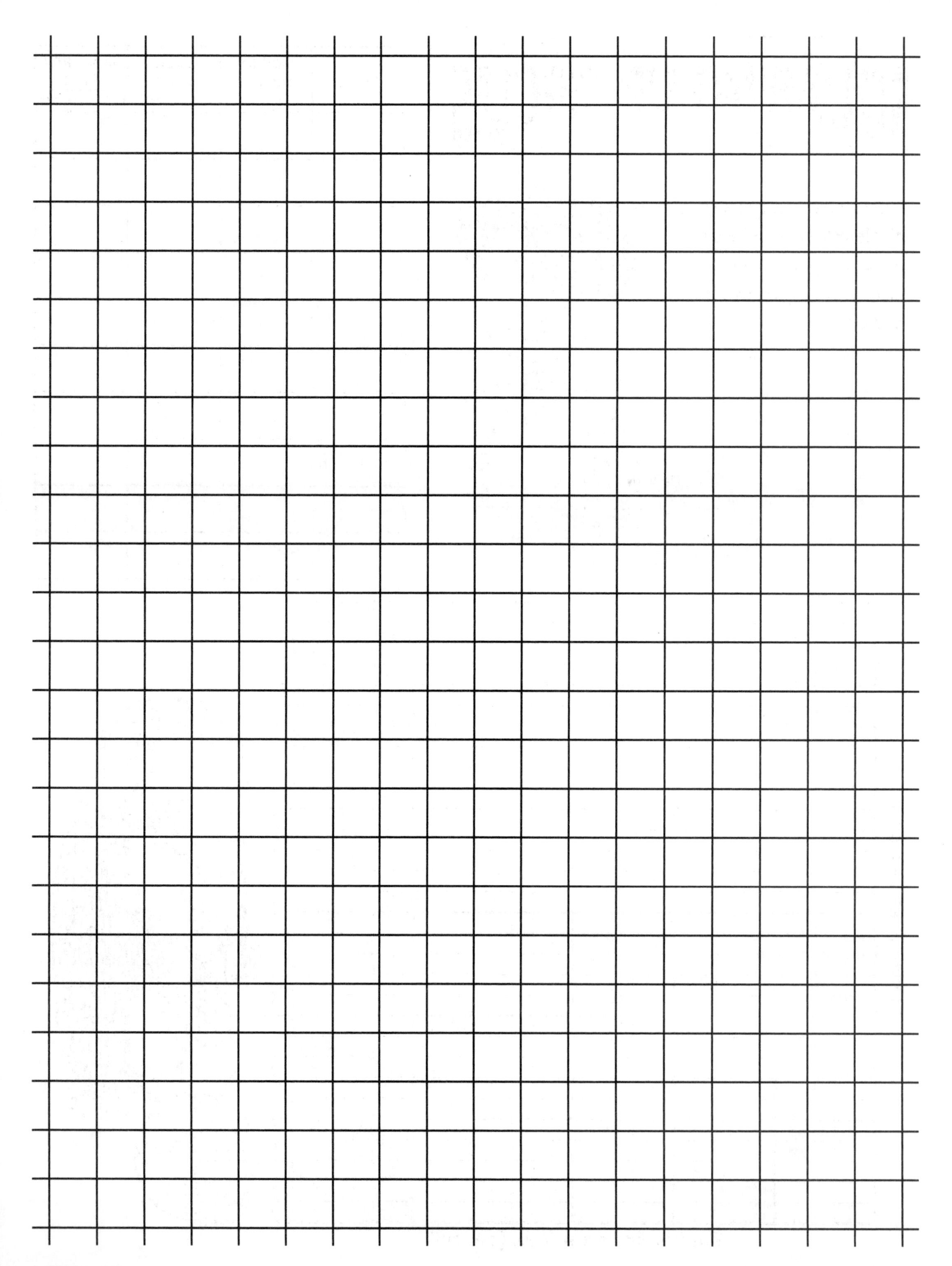

SQUARING UP CIRCLES

How does the area of a circle relate to the area of the square with a side the length of the radius?

Circle	Radius Square (sq. cm)	Circle Area (sq. cm)	Dec. Ratio Area / Radius Sq.

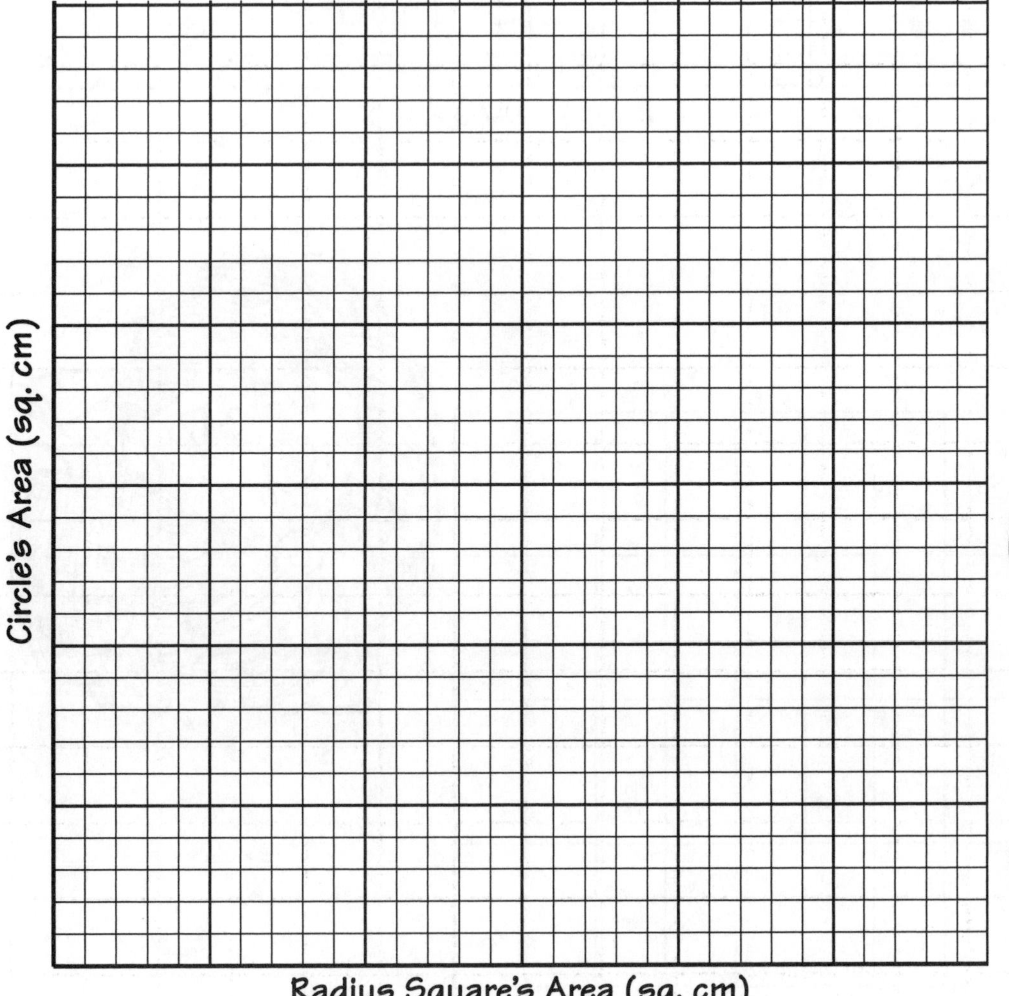

SQUARING UP CIRCLES

How does the area of a circle relate to the length of the circle's radius?

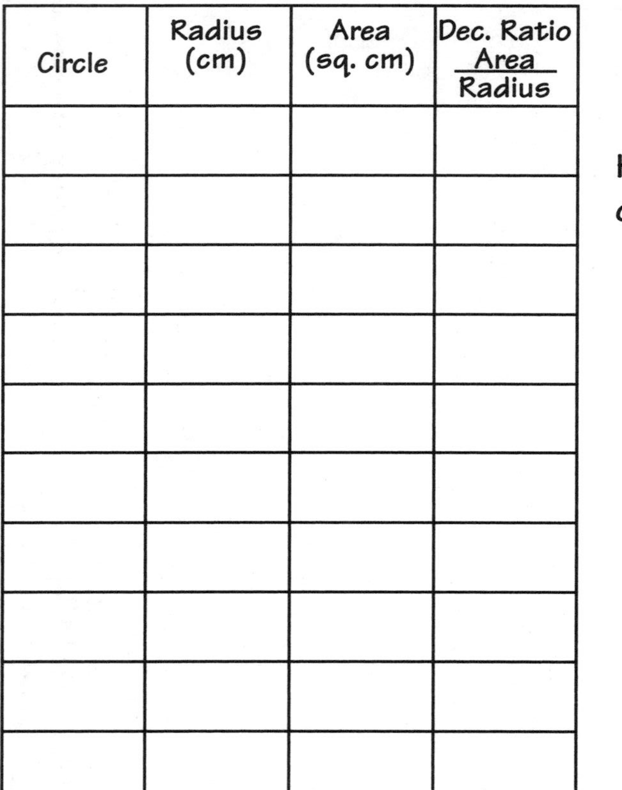

Circle	Radius (cm)	Area (sq. cm)	Dec. Ratio $\frac{\text{Area}}{\text{Radius}}$

Circle's Area (sq. cm)

Circle's Radius (cm)

GEOMETRY FORMULAS

Even when students have developed formulas for geometric relationships, they often have difficulty applying them in context. The difficulty arises because students develop the formulas by using known quantities to determine unknown quantities. With triangles students come to recognize the pattern that the area is one-half the product of the base and height. In none of their experiences have they known the area and had to determine the dimensions. By posing situations in which students need to work backwards to determine the unknown terms, their understanding is enriched and they develop their algebraic thinking.

The following pages provide experiences students might be asked to considered. Visual clues have been included with the problems so students will be encouraged to utilize the visual experiences they had developing the formulas and will apply them in finding solutions. In completing the pages students are often seen adding sketches to the illustrations to relate their thinking visually.

Students' understanding increases as they share solution strategies with each other. Most attack the problems in a numeric fashion. These students can be moved towards a more formal algebraic solution by observing a peer or teacher translating what they did into a more algebraic approach. Consider the two following problems:

If the length of a rectangle is twice its width and its perimeter is 24 inches, determine the length and width. Many students use a guess-and-check strategy to solve this problem. They guess the width, double it to determine the length, add the dimensions together, and double them to check if the guessed width gave them the correct answer. Some students during the process realize that a length is the same as two widths. This means the perimeter has a total of six widths. Algebraically their thinking could be symbolized:

Perimeter $= w + l + w + l$

since for this problem the perimeter is 24 and $l = 2w$, substitute

$$24 = w + 2w + w + 2w$$
$$24 = 6w$$
$$24 \div 6 = 6w \div 6$$
$$4 = w$$
$$l = 2w = 2(4) = 8$$

the width is 4 and the length is 8, twice as large.

The area of a triangle is 54 m^2 and has a base of 12 meters. What is its height? Most students explain their thinking by saying they doubled the area of the triangle to get the area of the corresponding parallelogram (108 m^2). Then they used the parallelogram formula ($b \bullet h$ = area) and think 12 times something is 108. They do the division ($108 \div 12 = 9$) to determine the missing factor. Solving this problem algebraically models what students have thought and demonstrates to them the power of algebra.

$$\text{Area} = (b \bullet h) \div 2$$
$$54 = (12h) \div 2$$
$$54 \bullet 2 = (12h) \div 2 \bullet 2$$
$$108 = 12h$$
$$108 \div 12 = 12h \div 12$$
$$9 = h$$

As students work a number of backward problems, they gain a thorough understanding of the formulas. As they are also exposed and encouraged to record what they think algebraically, they prepare themselves for formal algebraic skills.

Solution Key

Working Backwards with *Rectangles 1*

1. 50 cm^2
2. 4 cm
3. 4 in.
4. 40 cm^2
5. 3.5 yds.
6. 14 m
7. 27 $in.^2$
8. 5 in.
9. 9.5 cm
10. 54 $ft.^2$
11. 4 m
12. 15 ft.

Working Backwards with *Rectangles 2*

1. 14 ft.
2. 6 ft.
3. 37.5 ft.
4. 55.5...≅55 ft.
5. 40 ft.
6. 1 $mi.^2$
7. 1200 seats
8. 288 people
9. 3200 $ft.^2$
10. 108 $in.^2$
11. 6" x 8"
12. 8' x 10'
13. 3 mi. x 4 mi.
14. 12' x 14'
15. 30 yds. x 40 yds.

Working Backwards with *Parallelograms*

1. 50 cm^2
2. 4 cm
3. 18 $in.^2$
4. 24.5 cm^2
5. 4.5 yds.
6. 13.5 $ft.^2$
7. 48 $ft.^2$
8. 5.5 in.
9. 4 m
10. 37.5 $ft.^2$
11. 2.5 ft.
12. 5 cm

Working Backwards with *Triangles*

1. 144 cm^2
2. 8 cm
3. 5 in.
4. 25 $in.^2$
5. 4 yd.
6. 12 m
7. 42 2/3 $in.^2$
8. 6 in.
9. 10 cm
10. 16 1/3 $ft.^2$
11. 9 m
12. 6 ft.

Working Backwards with *The Pythagorean Theorem*

1. 175.74 ft.
2. 15 in.
3. 127.28 ft.
4. 113.18 yds.
5. 14.14 ft.
6. 60.81 ft.
7. 7.07 cm
8. 11.31 in.
9. 0.87 m
10. 300 ft.

Working Backwards with *Trapezoids*

1. 216 cm^2
2. 4 cm
3. 8 in.
4. 23.625 cm^2
5. 6 yd.
6. 7 m
7. 600 $in.^2$
8. 6 in.
9. 15 cm
10. 45 $ft.^2$
11. 7 m
12. 18 ft.

Working Backwards with *Circles*

1. 81.64 in.
2. 19.1 turns
3. 26.5 cm
4. 2.34 ft. = 28 in.
5. 150.72 in.
6. 63.59 $in.^2$
7. 30.96 $in.^2$
8. 22.57 ft.
9. 127.38 lbs.
10. 10.66 rotations

1

RECTANGLES

1. The length of the rectangle is twice its width.

   5 cm

   What is the area?

2. The area of the rectangle is $12cm^2$.

   3 cm

   What is the length?

3. The length of the rectangle is twice its width.

   Perimeter = 24 in.

   What is the width?

4. The rectangle's length is 2.5 times its width.

   4 cm

   What is the area?

5. The area of the rectangle is 21 yd^2.

   6 yd,

   What is the width?

6. The width of the rectangle is half its length.

   Perimeter = 42 m

   What is the length?

7. The width of the rectangle is a third of its length.

   9 in.

   What is the area?

8. The area of the rectangle is $48in^2$.

   9.6 in

   What is the width?

9. The length of the rectangle is three times its width.

   Perimeter = 76 cm

   What is the width?

10. The rectangle's width is 1.5 times its length.

    9 ft.

    What is the area?

11. The area of the rectangle is $10m^2$.

    2.5 m

    What is the length?

12. The length of the rectangle is 1.5 times its width.

    Perimeter = 75 ft.

    What is the width?

2

RECTANGLES

Given the area and one dimension, determine the missing dimension.

1. One hundred sixty-eight square feet of carpet has been ordered for a 12 foot wide room. How long should the room be?

2. A pattern requires 27 square feet of fabric. If the fabric is woven 4.5 feet wide, how many feet of length is required?

3. A gallon of paint will cover 300 square feet. What length of an 8 foot high wall can you cover with 1 gallon of paint?

4. A bag of fertilizer covers 10,000 square feet. If you start spreading the fertilizer by going from side-to-side on a 180 foot wide soccer field, how far will you get down the field with a bag?

5. A snowplow can clear 400,000 square feet an hour. If a typical airport runway is 10,000 feet long, how wide a path can the snowplow clear in an hour?

Given one dimension and its relation to the other dimension, determine the area.

6. How much area is in a farmer's wheat field if it is four times longer than it is wide, and the road along the field's width is one-half mile long?

7. In every section of the college stadium there are three times as many seats in each row as there are rows of seats. If you count 20 rows of seats as you climb to the top of the stadium, how many seats are in each section?

8. The band director wants a rectangular formation of the marching band that has twice as many people in each column as there are people in a row. If 12 people are in the first row, how many people can be in the formation?

9. A basketball court's width is one-half of its length. If there is space for an 80 foot long court, how many square feet will it cover?

10. The width of sheet cake sold at the bakery is three-fourths of the length. If a cake is 12 inches long, how many square inches of cake will there be?

Given the perimeter and the area of a rectangle, determine the dimensions.

11. A frame is made from a 28 inch length of trim and surrounds a 48 square inch picture. What are the dimensions of the picture?

12. The asphalt play area was surrounded by a 36 foot long fence. The asphalt was then finished with a gallon of sealer that covers 80 square feet. What are the dimensions of the play area.

13. A marine patrol guards the 14-mile perimeter of its 12 square mile rectangular command base. What are the dimensions of the base?

14. One hundred sixty-eight square feet of ceramic tile has been delivered to finish the floor of the addition. Fifty-two feet of trim has been ordered to go along the base of the four walls that surround the rectangular addition. What are the dimensions of the addition?

15. A quarter acre lot (1200 square yards) is surrounded by 140 yards of fence. What are the dimensions of the lot?

1. The base of the parallelogram is twice its height.

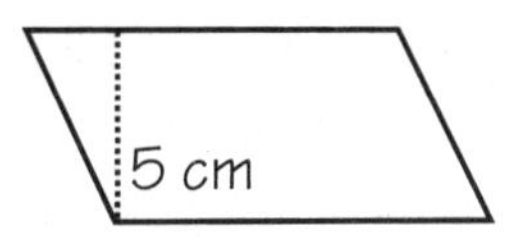

What is the area?

2. The area of the parallelogram is $12cm^2$.

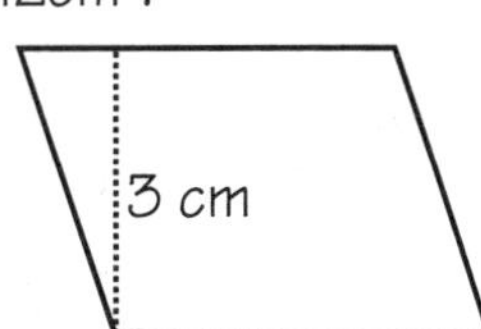

What is the length?

3. The base of the parallelogram is twice its height.

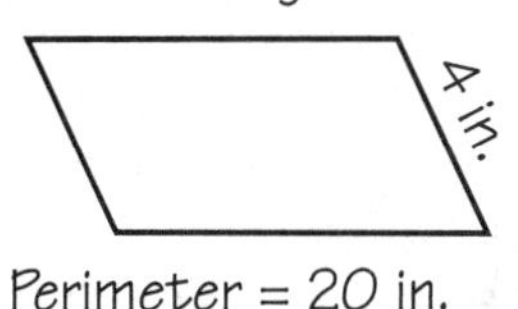

What is the area?

4. The parallelogram's height is twice its base's length.

What is the area?

5. The area of the parallelogram is 27 yd^2.

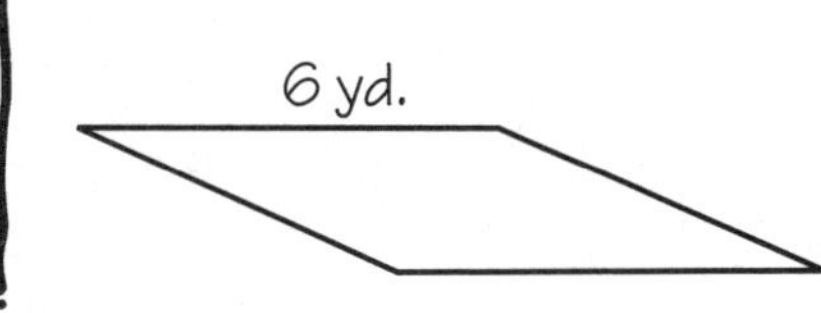

What is the height?

6. The height of the parallelogram is $1\frac{1}{2}$ times its base.

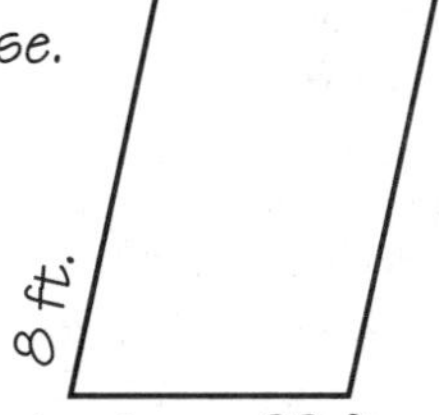

What is the area?

7. The base of the parallelogram is a third of its height.

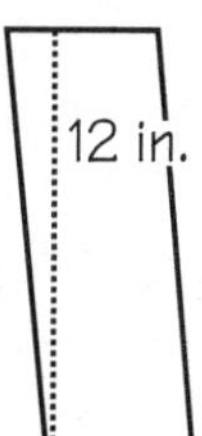
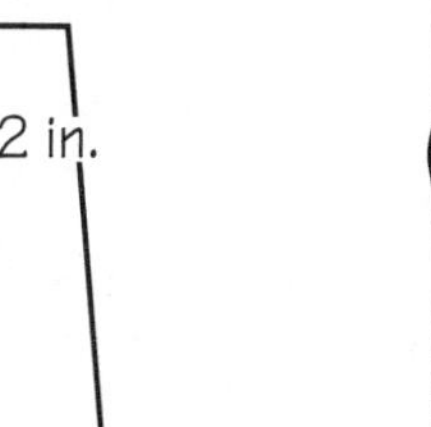

What is the area?

8. The area of the parallelogram is 44 in^2.

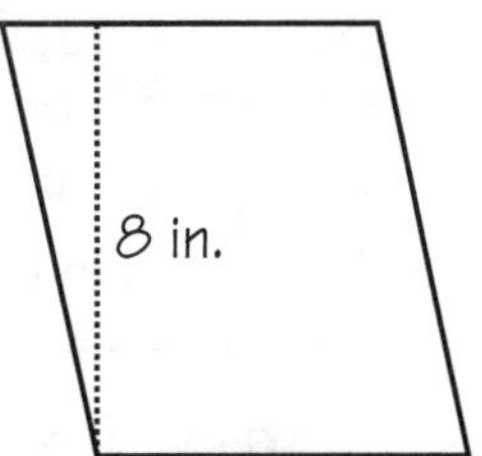
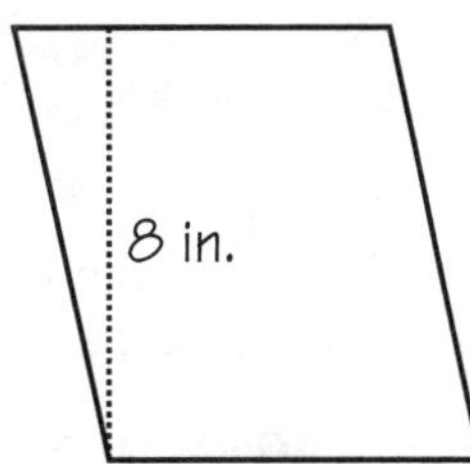

What is the base's length?

9. The height of the parallelogram is twice its base.

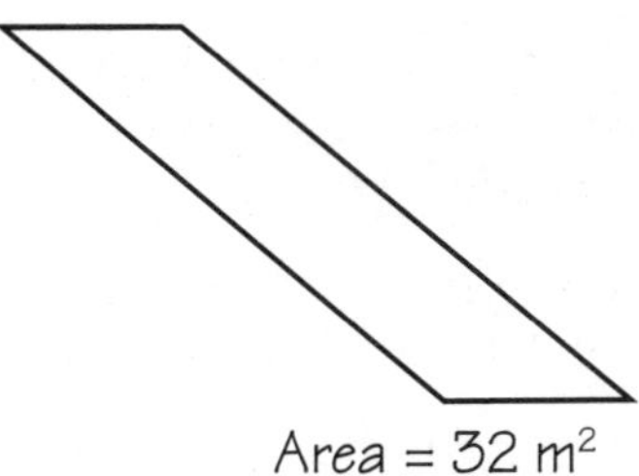

What is the base's length?

10. The parallelogram's height is $1\frac{1}{2}$ times the base's length.

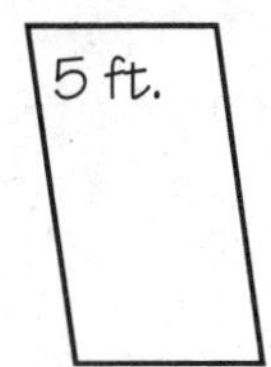

What is the area?

11. The area of the parallelogram is $10m^2$.

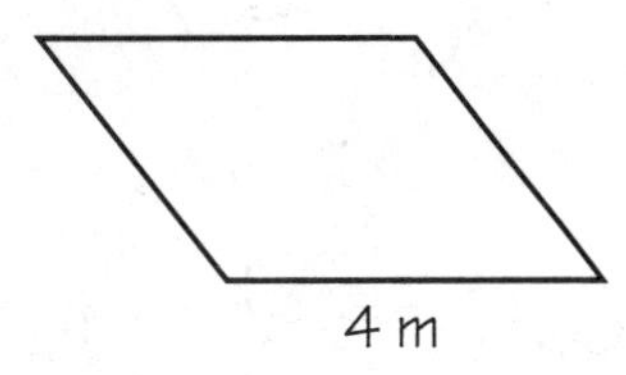

What is the height?

12. The base of the parallelogram is three times its height.

Area = 75 cm^2

What is the height?

1. The base of the triangle is twice its height.

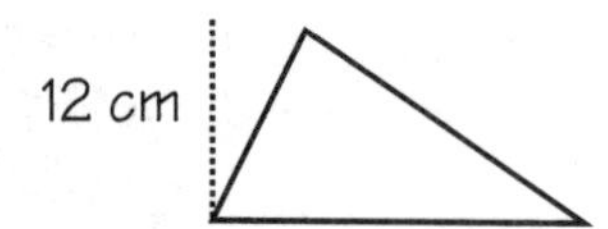

12 cm

What is the area?

2. The area of the triangleis $24 cm^2$.

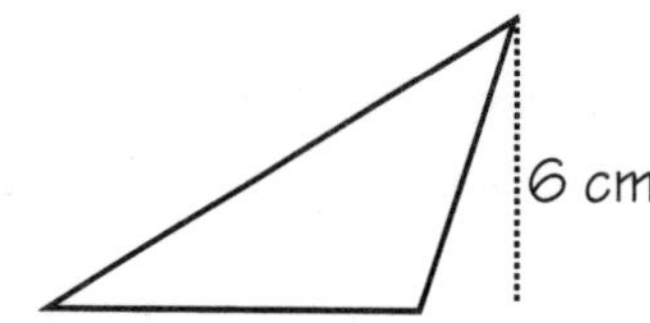

6 cm

What is the base's length?

3. The base of the triangle is twice its height.

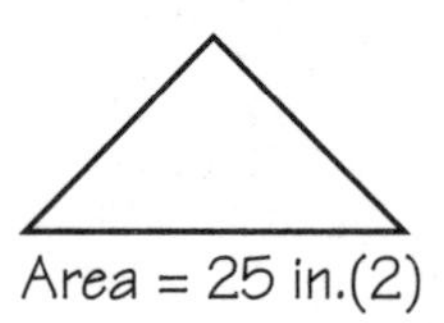

Area = 25 in.(2)

What is the height?

4. The triangle's height is twice its base's length.

5 cm

What is the area?

5. The area of the triangle is $18 yd^2$.

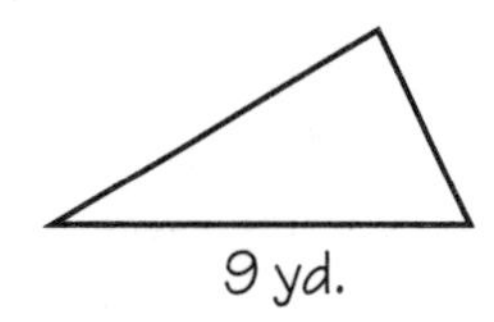

9 yd.

What is the height?

6. The height of the triangle is a third of its base.

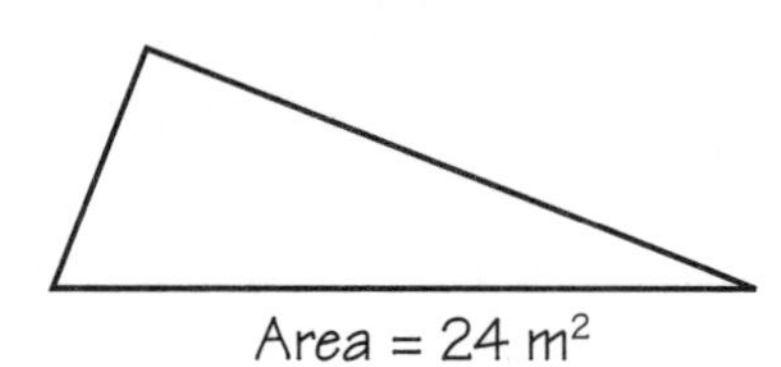

Area = $24 m^2$

What is the base's length?

7. The base of the triangle is a third of its height.

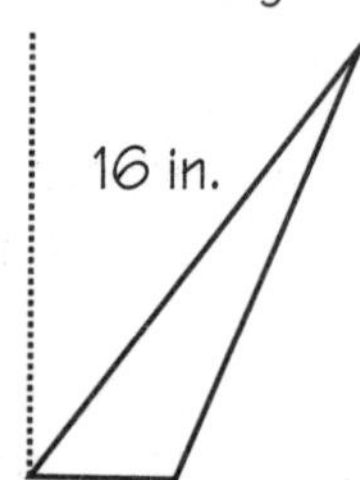

16 in.

What is the area?

8. The area of the triangle is $21 in^2$.

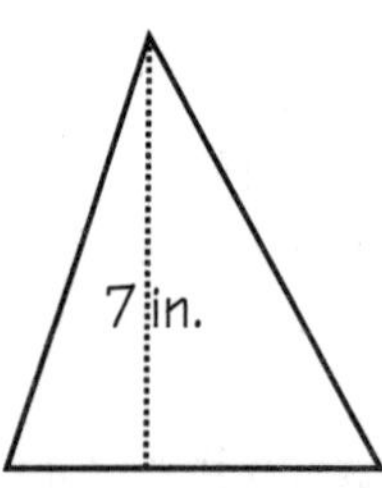

7 in.

What is the base's length?

9. The base of the triangle is one and a half times its height.

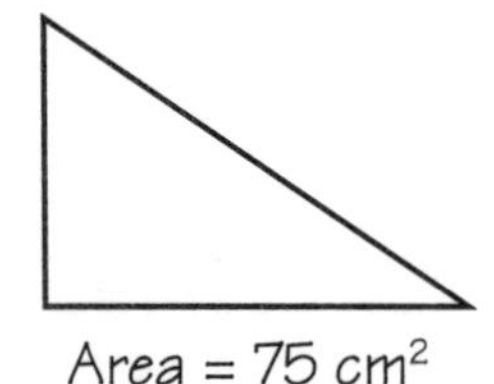

Area = $75 cm^2$

What is the height?

10. The triangle's height is $1\frac{1}{2}$ times the base's length.

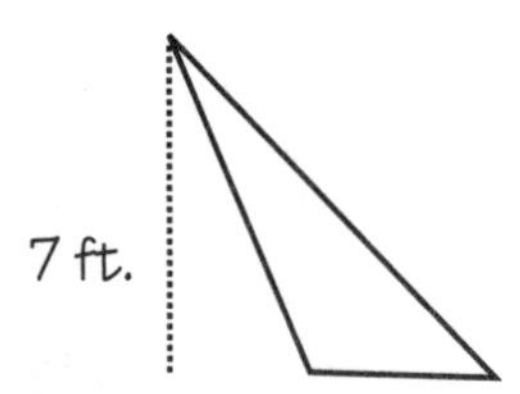

7 ft.

What is the area?

11. The area of the triangle is $54 m^2$.

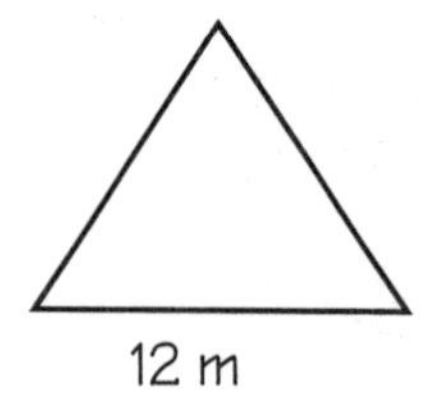

12 m

What is the height?

12. The height of the triangle is $1\frac{1}{2}$ times its base.

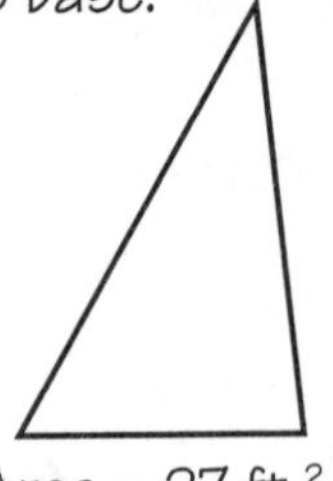

Area = $27 ft.^2$

What is the base's length?

The Pythagorean Theorem

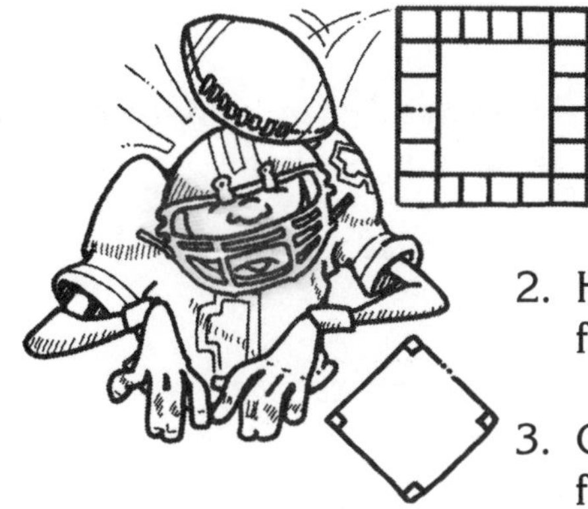

1. A 300 foot long and wide playing field is surrounded by a sidewalk. How much distance will you save if you take short-cut diagonally from the corners rather than walking on the sidewalk?

2. How long would the diagonals of a 9" x 12" picture frame be if it were square (had right angles)?

3. On a professional baseball diamond, there are 90 feet between the bases. How far is a throw from third to first base?

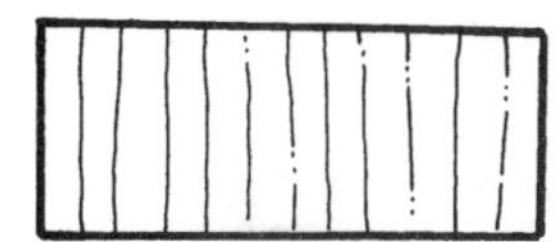

4. A football field is 100 yards long and 53 yards wide. How many yards long would a 100 yard kick off return be from one corner of the field diagonally to the opposite corner?

5. You are going to raise a banner for the school carnival. How long do you need to cut the rope for guy wires that need to be 10 feet from the ground and 10 feet from the pole?

6. In women's softball, the pitcher's mound is 43 feet from home plate. How far is it from home plate to first base?

7. For a project you need the diagonal fold of a square piece of origami paper to be 10 cm long. To what dimensions do you need to cut the square origami paper?

8. What are the dimensions of this tile pattern if it were cut from two eight-inch tiles?

9. What is the height of an equilateral triangle with an edge of one meter?

10. A kite is tethered to 500 feet of string. A girl is directly under the kite and stands 400 feet away from the person flying the kite. How high is the kite?

1. The top base of the trapezoid is the same as its height and half of the length of the bottom base.

12 cm

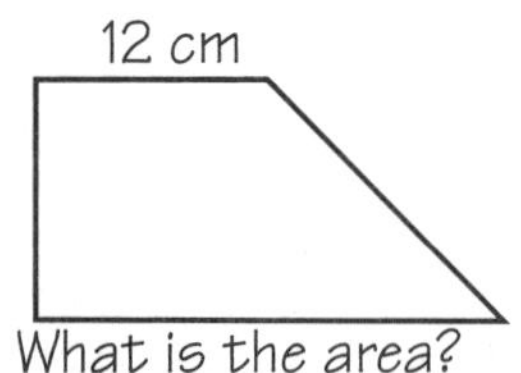

What is the area?

2. The area of the trapezoid is 36 cm ^{2}.

6 cm

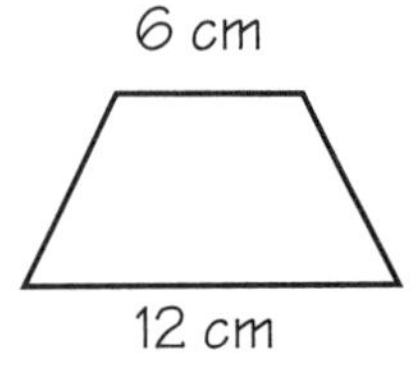

12 cm

What is the height?

3. The bottom base of the trapezoid is twice its height.

5 in.

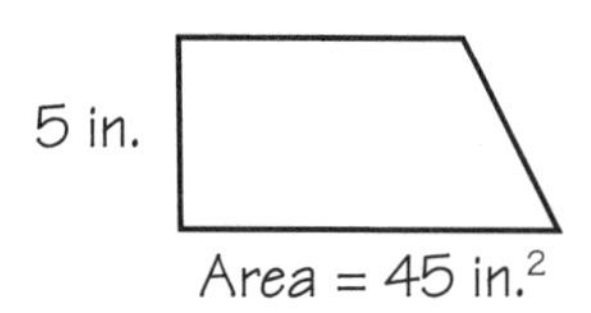

Area = 45 in.^{2}

What is the top base's length?

4. The height and top base are both $\frac{3}{4}$ of the bottom base's length.

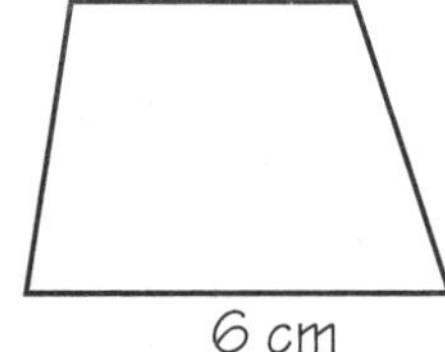

6 cm

What is the area?

5. The area of the trapezoid is 56 yd^2.

8 yd,

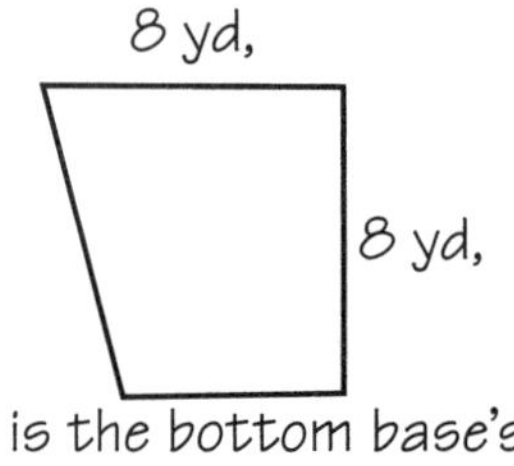

8 yd,

What is the bottom base's length?

6. The height of the trapezoid is a third of the bottom base's length.

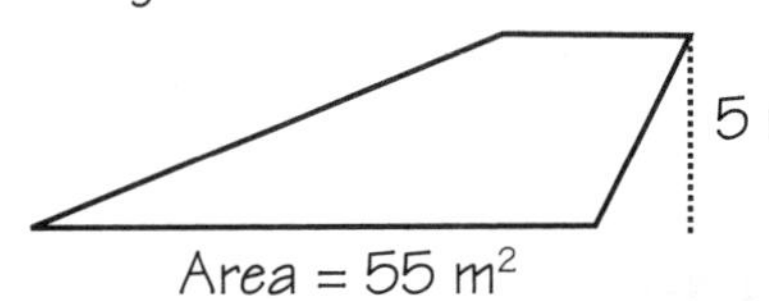

5 m

Area = 55 m^2

What is the top base's length?

7. The top base and height of the trapezoid are both 3 times the length of the bottom base.

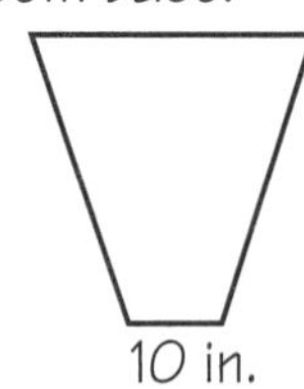

10 in.

What is the area?

8. The area of the trapezoid is 72 in^2.

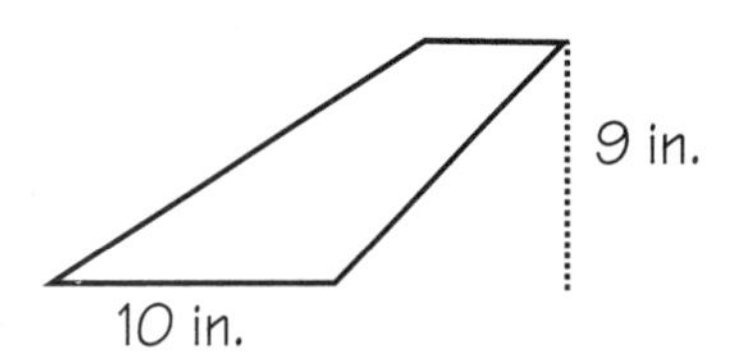

9 in.

10 in.

What is the top base's length?

9. The top base of the trapezoid is one and a half times its height.

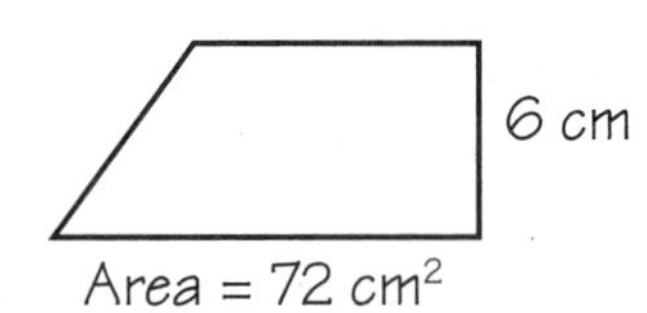

6 cm

Area = 72 cm^2

What is the bottom base's length?

10. The trapezoid's top base is half the height and the bottom base is twice the height.

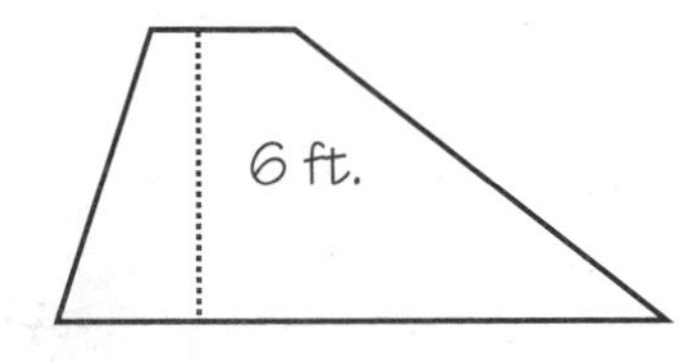

6 ft.

What is the area?

11. The area of the trapezoid is 49 m^2.

8 m

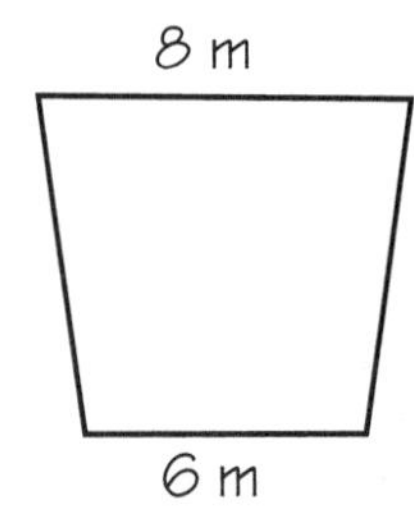

6 m

What is the height?

12. The height of the trapezoid is $1\frac{1}{2}$ times its top base.

6 ft.

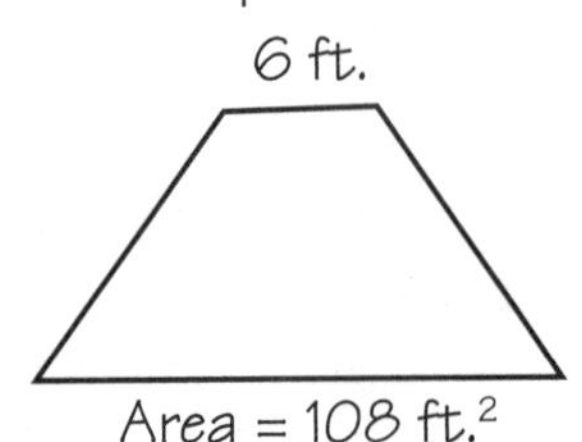

Area = 108 ft.^{2}

What is the bottom base's length?

CIRCLES

1. How far will a 26" diameter bicycle tire travel each time it turns?

2. How many turns will a 20" diameter bicycle tire make in a 100 foot race?

3. What is the diameter of a soccer ball that turns 12 rotations in 10 meters?

4. What is the diameter of a car tire that turns 180 times in 1320 feet ($\frac{1}{4}$ mile)?

5. How deep is a well that takes eight cranks on the six-inch diameter windlass to raise the bucket?

6. When a nine-inch diameter cream pie is thrown, how many square inches of the target will it cover?

7. How many square inches of paper remain from a 12-inch square after a 12- inch diameter circle is cut out of the square?

8. A gallon of paint covers 400 square feet. What is the biggest diameter circle of grass that can be covered as a target for a skydiver with a gallon of paint?

9. The groundskeeper needs to fertilize the infield of a 400-meter circular track. How many kilograms of fertilizer are required if 100 square meters are covered with each kilogram?

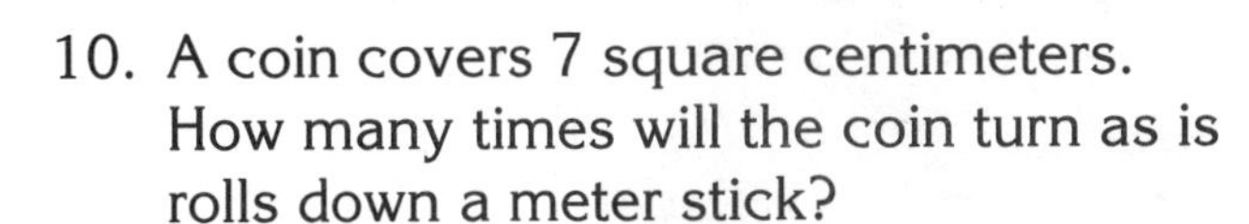

10. A coin covers 7 square centimeters. How many times will the coin turn as is rolls down a meter stick?

Topic
Properties of quadrilaterals with an emphasis on relationships

Key Question
How is this quadrilateral related to the others?

Learning Goals
Students will:
- determine the properties of a given quadrilateral and find all the shapes that also have those properties, and
- construct a graphic organizer showing the relationships among quadrilaterals.

Guiding Documents
Project 2061 Benchmark
- *Many objects can be described in terms of simple plane figures and solids. Shapes can be compared in terms of concepts such as parallel and perpendicular, congruence and similarity, and symmetry. Symmetry can be found by reflection, turns, or slides.*

*NCTM Standards 2000**
- *Identify, compare, and analyze attributes of two- and three-dimensional geometric shapes and develop vocabulary to describe the attributes*
- *Classify two- and three-dimensional shapes according to their properties and develop definitions of classes of shapes such as triangles and pyramids*

Math
Geometry and spatial sense
2-dimensional geometry

Integrated Processes
Observing
Collecting and recording data
Comparing and contrasting
Classifying
Relating

Materials
Card stock, preferably in colors
Scissors
Glue

Background Information
A quadrilateral is a polygon with four sides. Perhaps the most common one is the general quadrilateral. It has no special attributes—no congruent, parallel, or perpendicular sides—which may explain why it does not get the attention it deserves. The rest of the quadrilateral family has three main branches: trapezoids, parallelograms, and kites. Extending from these branches are more specialized family groupings, some of which are interrelated. For example, a square is a specialized form of a rectangle, as well a specialized form of a rhombus, but it is also part of the larger family of parallelograms. The following chart shows these relationships.

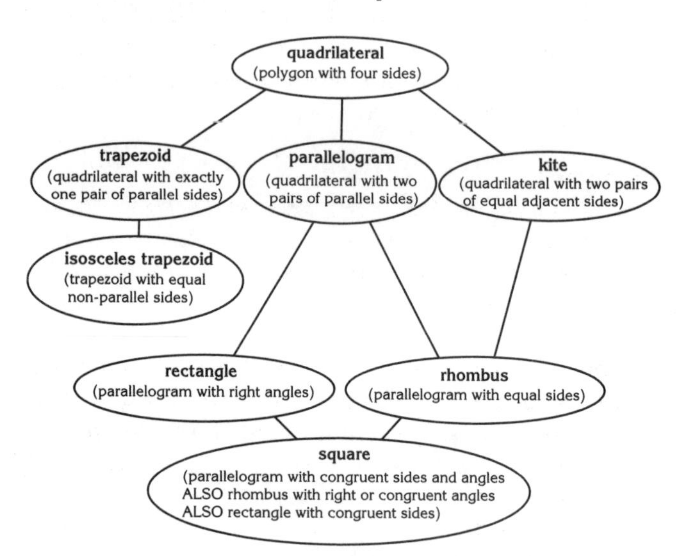

It should be noted that this is one of several ways in which mathematicians organize quadrilateral relationships, depending on how they define the individual shapes. Students should be encouraged to remain open to alternative diagrams they will likely encounter during their lifetimes.

Management
1. Copy the shapes page on card stock, preferably in colors. These shapes will be traced several times and later glued onto a graphic organizer. Use copy paper for the two graphic organizer pages. Copy the word bank page and cut in half or display the information on a piece of chart paper or on the chalkboard.

   While a recording page is included as an example of the information that should be gathered, you are encouraged to have students record the same information on a plain piece of paper folded in half. This option allows more freedom in organizing and presenting results and more tracing room as some of the records will not fit in the boxes provided. The activity page best meets needs when the shape to be scrutinized is a paralleogram or a kite because the other shapes are more evenly distributed between the *yes* and *no* boxes. Each chosen shape will need a separate recording page.

2. It is assumed students have previous experience with determining the properties of quadrilaterals. Advanced thinking is required of them here as they analyze how shapes are related, how they are alike and different.
3. You are encouraged to have students observe a good number of quadrilaterals as it will help cement their understanding of the properties and the relationships. In order not to exhaust the class, it is suggested one or two shapes be chosen each day for a week or so.
4. Cut a set of quadrilaterals ahead of time to introduce the activity. They can also be placed on an overhead projector when discussing properties and yes/no classifications.

Procedure

1. Put the 13 quadrilateral shapes on an overhead projector. (Alternatively, place the shapes on a table around which the class has gathered.) Pick up one shape and ask the *Key Question*: "How is this quadrilateral related to the others?" [It has four sides. (Entertain other observations about sides, angles, or parallel lines that are pertinent to that shape.)] Explain that students will be looking at relationships among quadrilaterals in this activity.
2. Distribute the shapes page and instruct students to cut out the 13 quadrilaterals.
3. Give students the word bank information and paper to record information. Pick one of the quadrilateral shapes so everyone starts with the same experience. Instruct the students to trace around the shape and record its properties, using the word bank page as a guide.
4. Have students share the properties they recorded and discuss as needed.
5. Direct students to examine the other 12 quadrilateral shapes. Those with the same properties as the chosen quadrilateral should be traced in a *yes* column and those which don't should be traced in a *no* column.
6. Discuss the results as a class. This is also the time to make sure everyone understands the process.

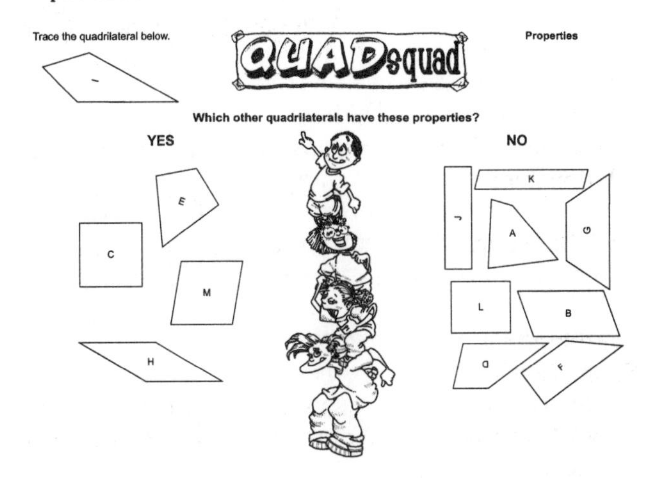

7. Repeat several times, each time using a different quadrilateral as the chosen shape.
8. Distribute the two graphic organizer pages and ask students to glue them together. Have students then place the 13 quadrilateral shapes in the appropriate places and glue them down.
9. Instruct students to pull out one of their earlier recording pages and explain how their *yes* shapes are connected to the graphic organizer. Repeat with other recording pages.

Discussion

1. What properties does a _______ (name a shape) have? (For example, a square has four equal sides. Both pairs of opposite sides are parallel to each other. Sides which meet—adjacent sides—are perpendicular. The square has equal angles.)
2. How could we check the angles? [measure with a corner of a paper or a protractor]
3. How do your *yes* shapes for ______ (name the letter of a quadrilateral shape) go with what is on the chart? (Example: For the kite, the chart pathway goes from the kite to the rhombus to the square and those are the same shapes that I have in the *yes* group.)
4. How could you then use your chart to tell *yes* answers? [Just name a shape letter, find it on the chart, and all the shapes that are linked below it will be in the *yes* column.]

Extension

Art: Family Portraits

Divide the class into eight groups, one for each of the quadrilateral families: general quadrilateral, trapezoid, isosceles trapezoid, parallelogram, rectangle, rhombus, square, and kite. Assign or have each group choose one of the families. Take kites for an example. Each kite group member will make a "people" portrait using one of the kite shapes as the head. As only two kite shapes are included in the activity, additional kites will likely need to be made.

The individual portraits can then be cut and grouped as a family sitting, framed by construction paper and a label. When the eight families are completed, display on the bulletin board in the pattern of the graphic organizer.

* Reprinted with permission from *Principles and Standards for School Mathematics,* 2000 by the National Council of Teachers of Mathematics. All rights reserved.

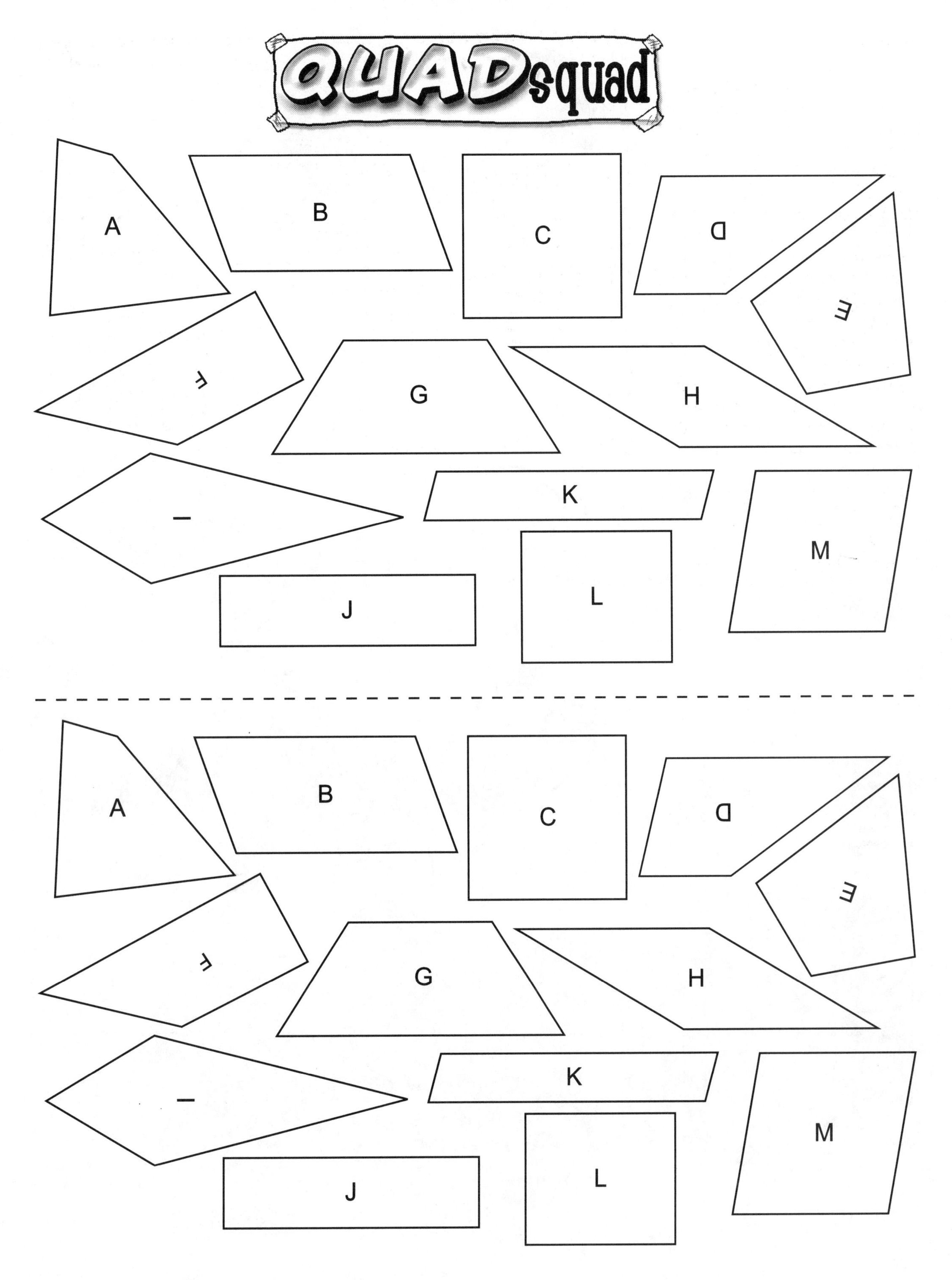
QUADsquad
A
B
C
D
E
F
G
H
I
K
M
J
L
A
B
C
D
E
F
G
H
I
K
M
J
L

Properties Word Bank

side ....line segment
angle ....two rays with a common endpoint
equal ....same measure
congruent ....same size and same shape
parallel ....will never intersect
perpendicular ....at right or 90° angles
opposite ....across from
adjacent ....next to

QUADsquad

Properties Word Bank

side ....line segment
angle ....two rays with a common endpoint
equal ....same measure
congruent ....same size and same shape
parallel ....will never intersect
perpendicular ....at right or 90° angles
opposite ....across from
adjacent ....next to

Are any angles congruent?
How many sides? How many angles?
Are any sides parallel?
Are any sides congruent?

Trace the quadrilateral below.

Properties

Which other quadrilaterals have these properties?

YES

NO

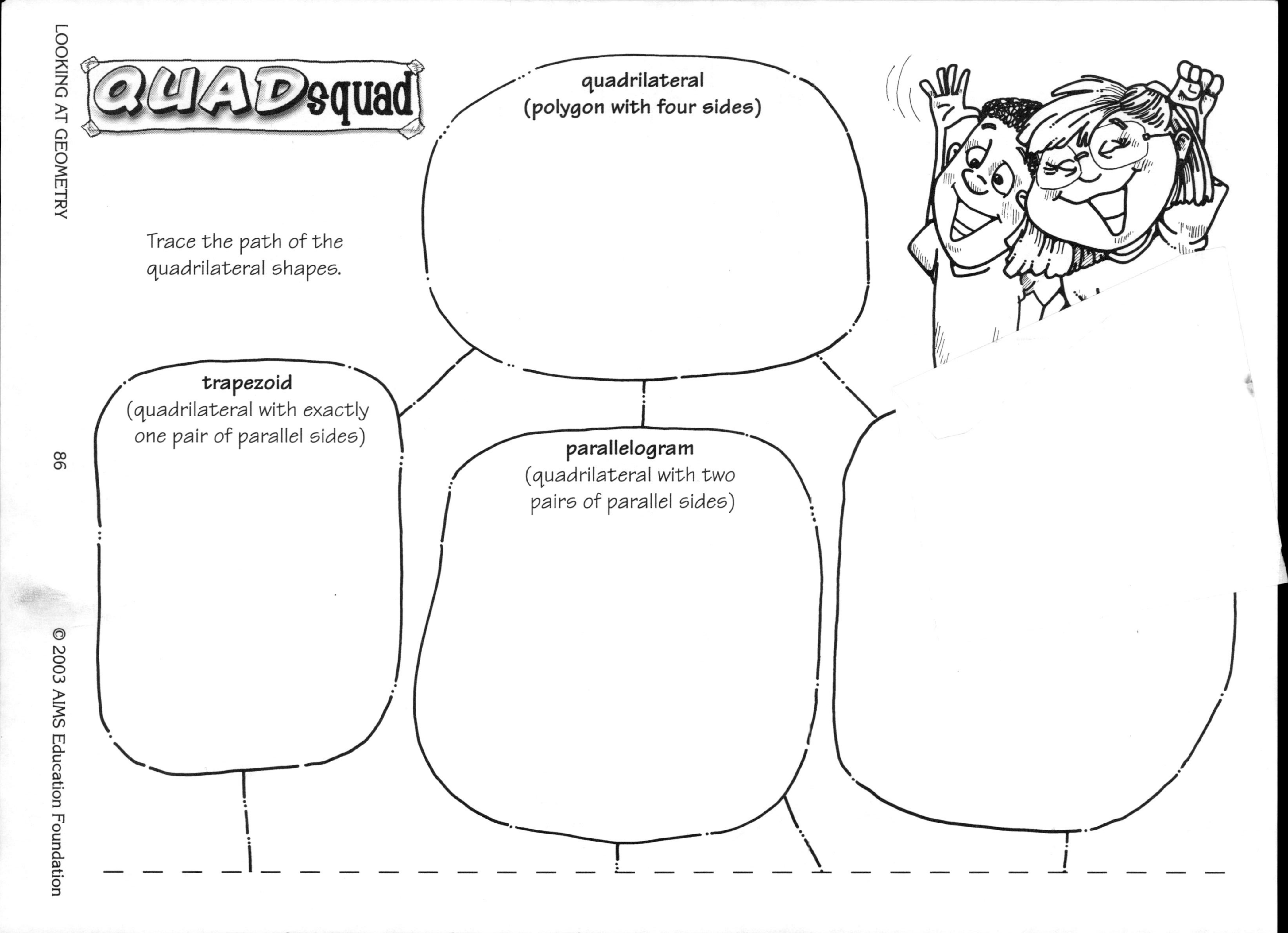
QUADsquad
Trace the path of the quadrilateral shapes.
quadrilateral
(polygon with four sides)
trapezoid
(quadrilateral with exactly one pair of parallel sides)
parallelogram
(quadrilateral with two pairs of parallel sides)

glue line
isosceles trapezoid
(trapezoid with equal
non-parallel sides)
rectangle
(parallelogram with right angles)
rhombus
(parallelogram with equal sides)
square
(parallelogram with congruent sides and angles)

Topic
Classifying geometric solids

Key Question
How are the geometric solids similar and different?

Learning Goals
Students will:
- sort and classify geometric solids by similar and distinguishing properties, and
- identify geometric solids by their properties.

Guiding Documents
Project 2061 Benchmarks
- *Some shapes have special properties: Triangular shapes tend to make structures rigid, and round shapes give the least possible boundary for a given amount of interior area. Shapes can match exactly or have the same shape in different sizes.*
- *Lines can be parallel, perpendicular, or oblique.*

*NCTM Standard 2000**
- *Precisely describe, classify, and understand relationships among types of two- and three-dimensional objects using their defining properties*

Math
Geometry
solids

Integrated Processes
Observing
Comparing and contrasting
Sorting
Classifying

Materials
3-D model sets (see *Management 1)*
Butcher paper
Markers

Background Information
A solid is the surface that surrounds a three-dimensional space. The solids found as models in sets can be sorted into five categories: spheres, cylinders, prisms, cones, and pyramids. All mathematicians agree on which category most precisely describes a solid. They differ in the definition of the categories. Some mathematicians focus on distinctives and develop exclusive definitions such as a cylinder has a base of a simple closed curve that is not a polygon and a prism has a base of a polygon. Other mathematicians focus on commonalties and are more inclusive defining a cylinder as having a simple closed curve as a base. The prism becomes a subset of that group having a polygon as the base's simple closed curve. The more inclusive definition encourages students to focus on similarities of attributes that align with similarity of volume formulas. For this series of investigations, the more inclusive definitions will be encouraged. With that in mind, except for the sphere and hemisphere, all of the solids in the model sets can be defined as cylinders or cones.

A cone is geometric solid including a base bounded by a simple closed curve and a lateral surface composed of line segments joining every point on the boundary of that curve to a common vertex. The circular cone comes to mind because it is most commonly used as an illustration of this category of solids. The triangular and square pyramids have all the qualifying properties and can be categorized as cones but rarely are because we name them with the more specific category of pyramid. A pyramid is a cone with a polygon as a base. Students will often correctly put the pointed cones into one category

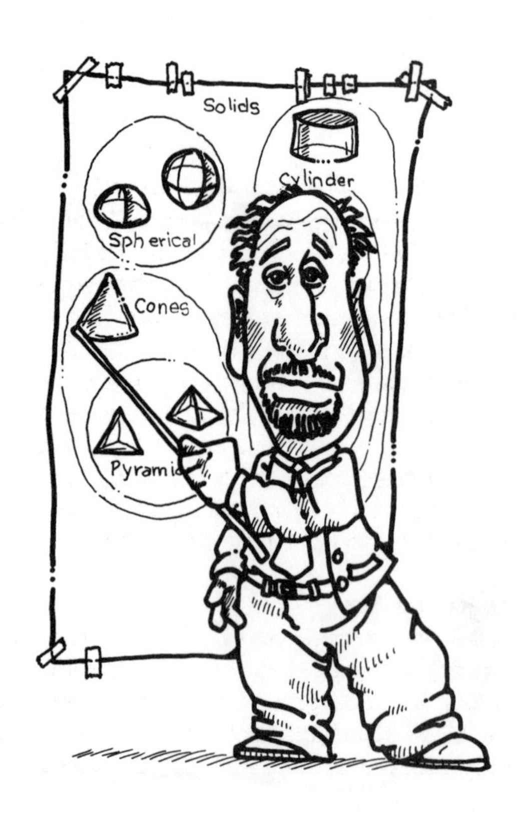

because of this obvious property. Likewise they are quick to recognize that the pyramids are different because of the flat sides.

A cylinder is composed of two congruent bases in parallel planes bounded by a simple closed curve and the lateral surface formed by line segments joining all the corresponding points on the boundaries. Again the circular cylinder is what most people identify with this definition. A prism is a cylinder with a polygon as a base. The hexagonal, triangular, and rectangular prisms are all cylinders but are named by the most specific definition of their properties. For this reason, a cube is not called a rectangular prism since cube is a more precise name.

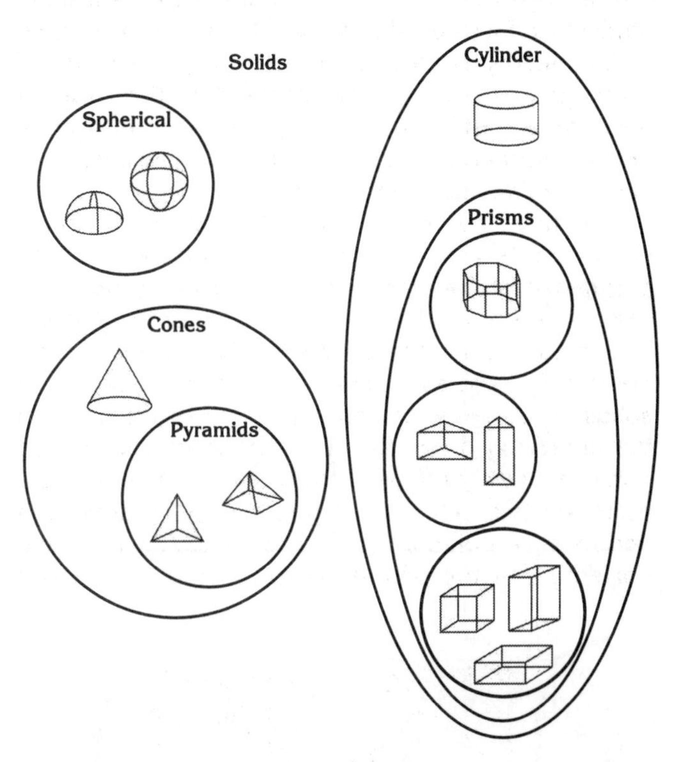

As students work with solid model sets, they will initially identify shapes by familiar objects and descriptive attributes. The cylinders might be called boxes and tubes, while "pointed shapes" might be used to name the cones. The emphasis should be to identify commonalties and distinguishing properties. As students identify these characteristics, the correct vocabulary can be introduced. The teacher should be aware as students learn the correct vocabulary, a child identifying a triangular pyramid as a triangular cone has recognized the common and distinguishing properties and may have simply forgotten the proper name.

Management

1. A manufactured set of 12 3-D models (Power Solids) can be obtained through AIMS. A set of solids for each group of four students is optimal for allowing all students to explore the solids. If only one set is available for the class, have the models out for reference and provide each group with a set of pictures to refer to and sort.
2. Butcher paper might be used to provide a sorting area and then have the groups use markers to record their sorting method. At the end of the lesson when students are familiar with properties and names of categories, each group can use the back side of the paper to resort and record the solids.

Procedure

1. Distribute the Power Solids and have students explore them, sorting them into groups by different properties. The students might be asked to record their sorting schema.
2. Have each group report to the class how they chose to sort the solids. Have a class discussion on how the solids were the same and different.
3. Have the students sort all the solids into groups based on the shape (circle, triangle, rectangle or square, hexagon) of their tops (the base with the hole in it).
4. As a class, discuss the similarities and differences within each group of solids and between the groups of solids.
5. Have the students select all the solids that come to a point opposite their tops. Inform the class that this group of solids is called cones.
6. Discuss similarities and differences of the three cones. When the students have identified that two of the cones have flat sides or polygon bases (bases with straight lines), inform them that these properties make them special cones called pyramids.
7. Have the students exclude the cones and consider the remaining solids. Have them determine and discuss what they all have in common that defines them as cylinders and what distinguishes them from each other making another special group called prisms.
8. Have the students resort the solids and record their relationships and properties.

Discussion

1. How are the solids in the set the same? [same height]
2. What similar properties did you use to sort the solids into groups? [pointed, spherical, shape of base]
3. What different properties distinguish the solids in each of your groups? [shape, size, curved or flat faces]
4. How are the solids based on a circle different from all the other solids? [curved surfaces]

SORTING SOLIDS

Topic
Surface areas of solids

Key Question
How can you construct a paper suit (net) for one of the solids and determine how much paper will be required?

Learning Goals
Students will:
- construct a net for a solid from the component pieces, and
- determine the surface area of a solid.

Guiding Documents
Project 2061 Benchmarks
- *Shapes on a sphere like the earth cannot be depicted on a flat surface without some distortion.*
- *Calculate the circumferences and areas of rectangles, triangles, and circles, and the volumes of rectangular solids.*

*NCTM Standards 2000**
- *Precisely describe, classify, and understand relationships among types of two- and three-dimensional objects using their defining properties*
- *Use two-dimensional representations of three-dimensional objects to visualize and solve problems such as those involving surface area and volume*

Math
Geometry
  solids
    nets
  surface area

Integrated Processes
Observing
Comparing and contrasting
Applying

Materials
3-D model sets (see *Management 1*)
Centimeter-grid paper
Scissors
Rulers
Drawing compass

Background Information
The pattern of components for composing a solid is called a net. For all the Power Solids except for the sphere and hemisphere, this can be done with a combination of simple closed curves (polygons, circles) from a sheet of paper. If aligned correctly on a paper and cut out, the net can be folded to completely cover the solid.

Students progress through three methods of construction when given a solid or its dimensions and they are asked to make its net. The most direct method is to place the solid on a piece of paper and trace the base. After rolling the solid onto an adjacent surface, while keeping the shared edge aligned, students trace a second surface. Continuing this technique until all the surfaces have been traced can produce a net. When the students cut out the perimeter of the net, they can assemble it to confirm if it covers the surface of the solid. All the nets of a given solid will contain the same components, but they may be placed in a variety of alignments.

When students can visualize what shapes are required and how they align, they are ready to move to the second method of construction. At this level, students measure the solid and use their measurements to draw the net. The solids with rectangular faces are most easily drawn because all the edges meet at right angles. Compasses are required for the circular cylinder. Making the net for the circular cone from measurements is often beyond the understanding of students. They do not recognize that the length of the lateral surface is the radius of the circle from which the lateral surface is cut, while the arc is the circumference of the cone's circular base.

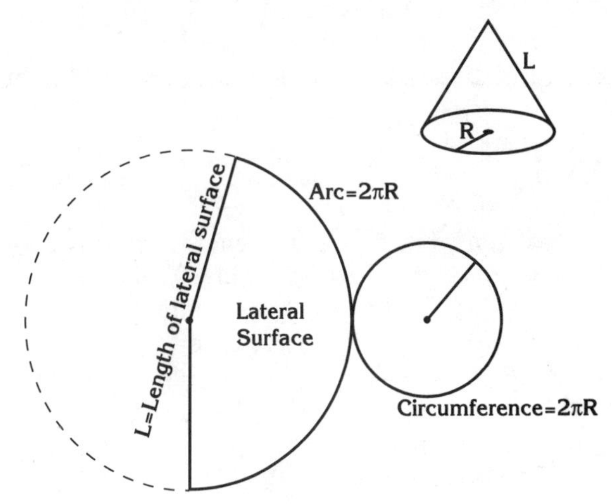

Triangular faces can be drawn using a compass to swing the lengths of two edges from the endpoints of the third until they intersect. Another practical solution is to draw a line perpendicular to the base of each triangle at the base's midpoint. Since all of the triangular faces are isosceles or equilateral triangles, the perpendicular bisector is the height of the triangle and is easily measured to define the triangle.

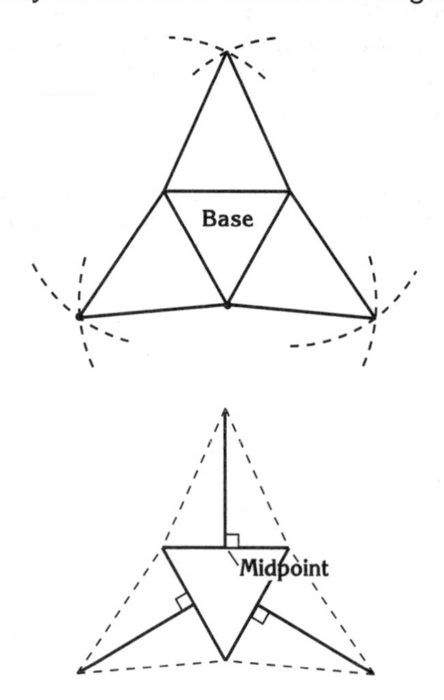

The most abstract way at approaching nets is to know that all the solids are based on a length of 5 cm (2 in). All the solids are 5 cm high. The edges or diameters of all the bases are 5 cm or half of that, 2.5 cm. Students generally do well with all but the cones and pyramids. A clear understanding of the Pythagorean relationship is required to determine the length of the lateral surfaces.

The net construction process can develop a better understanding of the measurement of the surface area. Intuitively it establishes the concept of surface area as a two-dimensional situation. Spatially, the construction process encourages students to visualize the component surfaces and their relationship to one another in both two and three dimensions. By using grid paper, the numeric calculation of area is encouraged.

Students who trace the net from the solid can be encouraged to count square area on the grid of the net. The patterns of numbers can be used to develop the idea that multiplying the length times the width is always part of an area formula. With rectangular regions, this pattern is most evident. Encouraging students to cut the triangular faces and reorganize the pieces into a rectangle provides a bridge for developing a formula for the triangle (refer to *Paper Cutting Geometry*). The experience of counting the area using the grid on the net and then relating it with the dimensions of the face provides an understanding of an area formula as a description of the pattern of the relationship.

As students become confident with area formulas, they can be encouraged to predict the area of different faces using the measurements. They can confirm their computations by counting the areas of the regions.

Management

1. A manufactured set of 12 3-D models (Power Solids) can be obtained through AIMS. A set of Power Solids for each group of four students is optimal for allowing all students to explore the solids. If only one set is available for the class, each group of students can use one of the solids at a time and share them around the class.
2. For accurate communication, it is necessary for the students to be familiar with the properties and names of the solids. Consider doing *Sorting Solids* before beginning this activity.
3. Before starting this investigation, determine if *Level One* or *Level Two* is most appropriate for your students and consider only doing that level. If you feel it is appropriate for students to work through both levels, a few solids should be selected to be done at each level.

Procedure

Level One

1. Discuss the *Key Question* with the class and ask students to suggest how nets might be made.
2. Provide the students with a set of Power Solids and grid paper and have them trace the solids to construct nets.
3. Have students count and record the area of the solid's surface along with its dimensions.
4. Discuss what observations were made about the nets of the solids and what patterns related the dimensions of a face to its area.

Level Two

1. Have student discuss techniques they could use to construct different shapes from dimensions (rectangle, triangle, circle). If the students appear to have difficulty, they should receive some instruction on construction techniques using a compass and ruler.
2. Provide a set of Power Solids to each group and tools and materials for drawing nets to each student. If there is only one set for the class, place each of the solids at different locations around the room so students can go to the various locations to make measurements.
3. Direct the students to take the measurements they need and record them. Using the dimensions, direct them to construct a net. When they have

constructed their nets, encourage them to check to see how well they cover the solid.

4. Have the students use the dimensions of their nets to calculate the area of each surface of the net. Instruct them to record their calculations and then to count the area to see if they are correct.

Discussion

1. How many surfaces does each type of solid have?
2. How many differently shaped surfaces does each solid have? [Excluding the sphere, hemisphere, and cube (1, square), all the solids have two different shaped surfaces.]
3. What connects all but the curved surfaces? [edge, line]
4. What different shapes make up all the nets? [rectangle, square, triangle, hexagon, circle or part of a circle]
5. How many different nets could you make for each solid?
6. What relationship do you see between the dimensions of a shape and its area that would help you know how much material you would need to construct the shape? [length x width, rectangle; (l•w)/2 = (w/2)• l = (l/2)• w, triangle]

Extension

Gather or have students bring in small examples of solids. Have the students take measurements of the samples and construct nets for them and calculate the surface area of each sample.

* Reprinted with permission from *Principles and Standards for School Mathematics,* 2000 by the National Council of Teachers of Mathematics. All rights reserved.

Suits for Solids

____
Type of Solid

Shape of Surface	Dimensions	Area of Surface	× Number	= Total SA
			Surface Area of Solid	

____
Type of Solid

Shape of Surface	Dimensions	Area of Surface	× Number	= Total SA
			Surface Area of Solid	

____
Type of Solid

Shape of Surface	Dimensions	Area of Surface	× Number	= Total SA
			Surface Area of Solid	

____
Type of Solid

Shape of Surface	Dimensions	Area of Surface	× Number	= Total SA
			Surface Area of Solid	

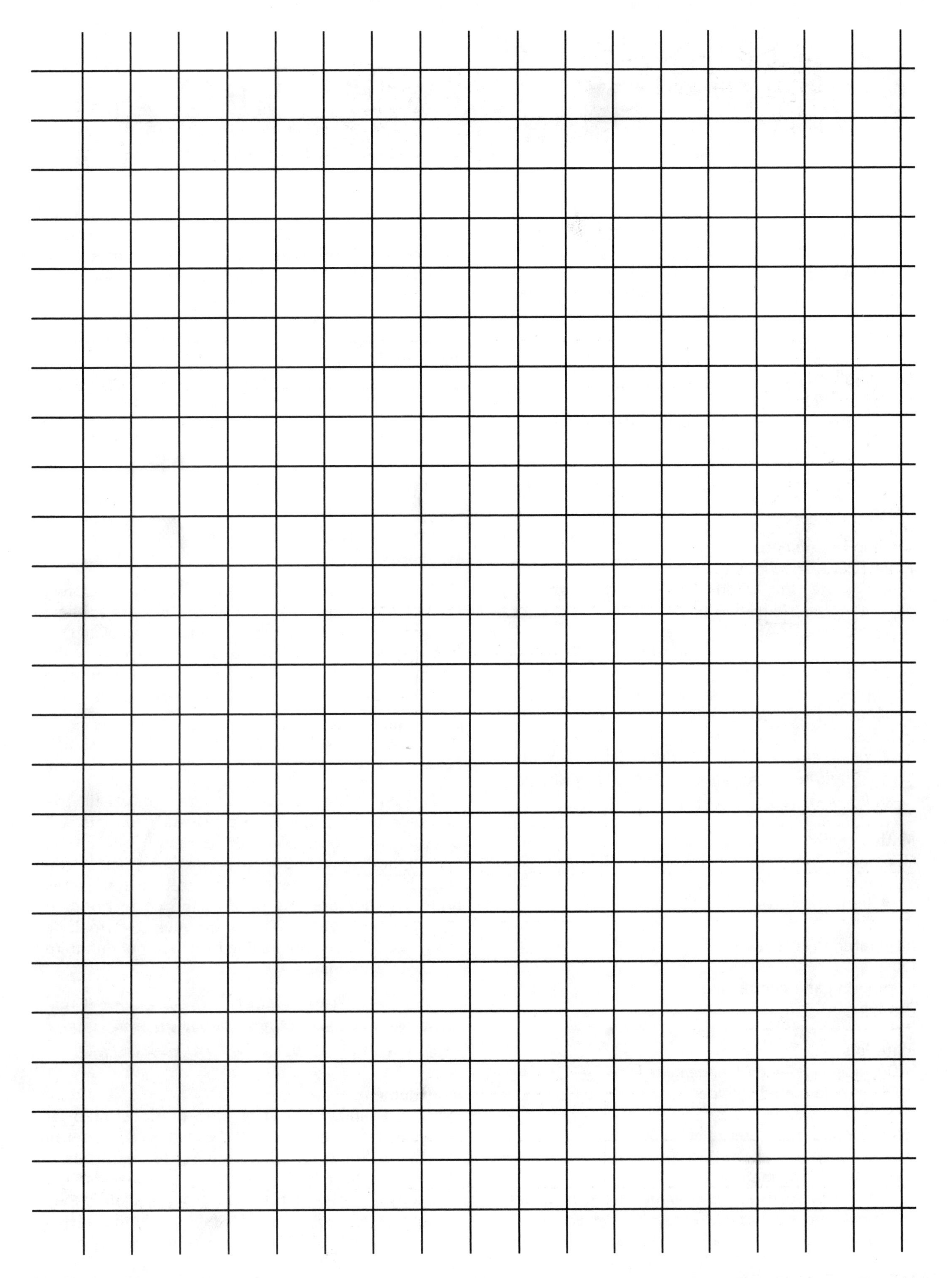

Topic
Volume

Key Question
How do the volumes of a cylinder (or prism) and cone (or pyramid) with the same linear dimensions compare?

Learning Goals
Students will learn:
- how the volumes of different solids are related,
- the meaning of the each expression in a volume formula, and
- to develop formulas to calculate volume of selected solids.

Guiding Documents
Project 2061 Benchmark
- *Calculate the circumferences and areas of rectangles, triangles, and circles, and the volumes of rectangular solids.*

*NCTM Standards 2000**
- *Understand relationships among the angles, side lengths, perimeters, areas, and volumes of similar objects*
- *Develop strategies to determine the surface area and volume of selected prisms, pyramids, and cylinders*

Math
Geometry
solids
volume formulas

Integrated Processes
Observing
Comparing and contrasting
Applying

Materials
3-D model sets (see *Management 1)*
1000 mL graduated cylinder
or
Liter box and millimeter ruler

Background Information
Geometric solids are three-dimensional by nature. To calculate the volume of any solid, measurements of its length, width, and height are required.

Power Solids provide an excellent opportunity for students to develop and confirm formulas for calculating volumes. The dimensions of all the solids are based on a standard length of 5 cm. Some of the smaller solids use a dimension of half the standard length (2.5 cm). This keeps the calculations simple and makes relationships between shapes more evident since comparisons can be made directly between the solids.

The cylinder and prisms should be the initial introduction for students. If students understand the concept of calculating the volume of a rectangular prism (box) as the product of the length, width, and height, the three rectangular prisms can be calculated easily:

Cube $V = b \bullet h$ $V = l \bullet w \bullet h$ $V = 5 \bullet 5 \bullet 5 = 125\ cm^3$
Large Rectangular Prism $V = b \bullet h$ $V = l \bullet w \bullet h$ $V = 5 \bullet 2.5 \bullet 5 = 62.5\ cm^3$
Small Rectangular Prism $V = b \bullet h$ $V = l \bullet w \bullet h$ $V = 2.5 \bullet 2.5 \bullet 5 = 31.25\ cm^3$

Students can verify their calculations by measuring the volume by displacement. From their experience, students can generalize that volume is calculated by knowing the area of the base (b) and multiplying it by the height (h) of the solid. Calculation and verification of the remaining prisms will confirm this generalization.

Cylinder $V = b \bullet h$ $V = 2\pi \bullet h$ $V = 3.14(2.52) \bullet 5 = 98.1\ cm^3$
Triangular Prism $V = b \bullet h$ $V = (.5l \bullet w) \bullet h$ $V = 0.5(5)(4.3) \bullet 5 = 53.8\ cm^3$
Hexagonal Prism $V = b \bullet h$ $V = (1.5s \bullet w) \bullet h$ $V = 1.5(2.5)(4.3) \bullet 5 = 80.6\ cm^3$

By using the cones and pyramids to fill the corresponding cylinder and prisms, students will observe that one cone or pyramid will fill its corresponding cylinder or prism one-third of the way. In this way, they will generalize that cones and pyramids are one-third of the corresponding cylinders and prisms. The basic volume formula is easily modified by multiplying by a factor of one-third for cones and prisms.

Cone $V = \frac{1}{3}(b \bullet h)$ $V = 1/3(2\pi \bullet h)$ $V = \frac{1}{3}(3.14(2.52) \bullet 5) = 98.1\ cm^3$
Triangular Pyramid $V = \frac{1}{3}(b \bullet h)$ $V = 1/3(.5l \bullet w) \bullet h)$ $V = \frac{1}{3}(0.5(5)(4.3) \bullet 5) = 53.8\ cm^3$
Square Pyramid $V = \frac{1}{3}(b \bullet h)$ $V = 1/3(l \bullet w \bullet h)$ $V = \frac{1}{3}(5 \bullet 5 \bullet 5) = 41.7\ cm^3$

Management
1. A manufactured set of 12 3-D models (Power Solids) can be obtained through AIMS. A set of Power Solids for each group of four students is optimal for allowing all students to measure the solids. If fewer sets of solids are available, they can be placed at central locations where students can use them one at a time as needed.

2. For accurate communication, it is necessary for the students to be familiar with the properties and names of the solids. Consider doing *Sorting Solids* before beginning this activity.
3. The activity is written for students to use the exterior dimensions to calculate volumes and check the accuracy of the calculations by displacement. This method provides simple numbers with which to work and seems to produce less error than using interior dimensions.

   The volume calculations may be checked by filling the interior with water, salt, sand, or even birdseed. The students will need to accurately measure the interior dimensions. This method tends to produce greater error with students due to spillage, surface tension, or other variables.
4. If large graduated cylinders are not available, liter boxes and millimeter rulers can be substituted to measure displacement. Each millimeter of rise in a liter box has a volume of 10 milliliters or 10 cubic centimeters.

Procedure

Cylinders and Prisms

1. Have the class discuss how they could measure and calculate the volume of box (rectangular prism). Elicit the response of multiplying the length, width, and height. If students do not have this understanding, the teacher should develop it with the class in another way before moving on with this investigation.
2. Instruct the students to measure exterior dimensions of the three rectangular prisms (boxes) from the set of Power Solids and calculate and record the dimensions and volume.
3. Have the students use a graduated cylinder of water and determine the volume of each of the boxes by displacement. They
   a. fill the graduated cylinder to 800 mL (800 cm^3),
   b. with the lid on the box, float it on the water,
   c. insert a pencil through the hole in the lid and push down on the bottom of the box,
   d. push the box into the water until the water level is even with the top of the box, and
   e. measure and record the rise in the water in cm^3 (mL).

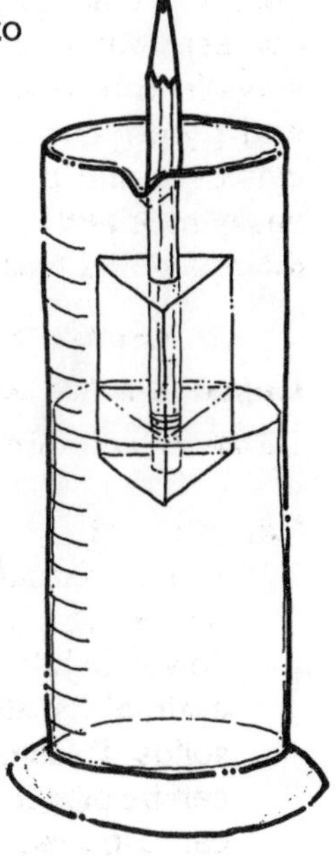

4. Have the class discuss the accuracy of measuring the volume of the solids using displacement.
5. Point out to the students that they determined the area of the base of each box, and then multiplied the area by the height of the box. Have them discuss how they might write out this generalization in mathematical terms. [$A = (l \bullet w) \bullet h = b \bullet h$]
6. Confirm that the class can determine the area of a triangle, hexagon, and circle. Then have them determine and record the area of the bases of the cylinder and prisms and calculate the volumes of those solids.
7. As the students measure, have them record the volume of the cylinder and prisms by displacement to check the accuracy of their calculations.

Cones and Pyramids

8. Instruct the students to match the cube, large triangular prism, and cylinder with their corresponding pyramid or cone. Ask them to describe each pair's similarities and differences.
9. Direct students to predict how the volume of each pyramid or cone compares to its corresponding prism or cylinder.
10. Tell students to fill each pyramid or cone with water and pour it into the corresponding prism or cylinder. Instruct them to move the cylinder and prisms next to each other and visually compare the volumes by seeing how much each pyramid or cone has filled up its corresponding solid.
11. By pouring more cone-fulls of water into the solids, have the students determine how many cones it takes to fill a cylinder with the same dimensions.
12. Have the students discuss how they might determine the volume of a cone since they know how to find the volume of a cylinder. Have them write their generalization as a formula. [$V = \frac{1}{3}(b \bullet h)$]
13. Using their formula and their calculations for the volume of the prisms and cylinder, have the students determine the volumes of the cone and pyramid, and then check them by displacement.

Discussion

1. How did the volume determined by calculations compare with the volume determined by displacement? (Answers may vary, but the results should be similar.)
2. How is the cone (pyramid) similar and different from its corresponding cylinder (prism)? [same base and height; lateral surface comes to a point, not to a face congruent to base]
3. What do you notice about the water level in the cylinder and prisms when you have poured one cone-full or pyramid-full of water into them? [all are about $\frac{1}{3}$ full]

4. How many cone-fulls or pyramid-fulls does it take to fill a corresponding cylinder or prism of the same dimensions? [about 3]
5. How does the volume of a cone or pyramid compare to the volume of a corresponding cylinder or prism of the same dimensions? [$\frac{1}{3}$]
6. Describe how you could use the relationships of the volume of a cone (pyramid) to a cylinder (prism) to determine the volume of the cone. How could you describe this method with a mathematical number sentence, an equation? [Find the volume of the cylinder and divide by 3; $V = (b \bullet h) \div 3 = \frac{1}{3}(b \bullet h)$.]

Extensions

1. Gather or have students bring small examples of solids. Have the students take measurements of the samples and calculate their volumes and then check them by displacement or filling. Have them use interior dimensions if they check by filling and exterior dimensions if they check by displacement.
2. Have students pour water between the cylinder, cone, hemisphere, and sphere to establish their volume relationships. Students will discover that the cone and the hemisphere have the same volume, one-third of the cylinder. The sphere can be filled with two cones or hemispheres making it two-thirds of the cylinder. If the diameter (2r) is considered the length, width, and height of the sphere, the equation students write for the sphere is based on the cylinder: $V = \frac{2}{3}(b \bullet h) = \frac{2}{3}(\pi r^2 \bullet 2r) = \frac{2}{3}(2\pi r^3) = \frac{4}{3}(\pi r^3)$

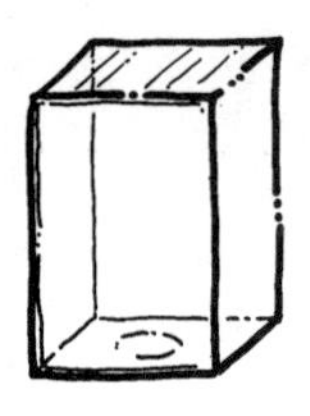
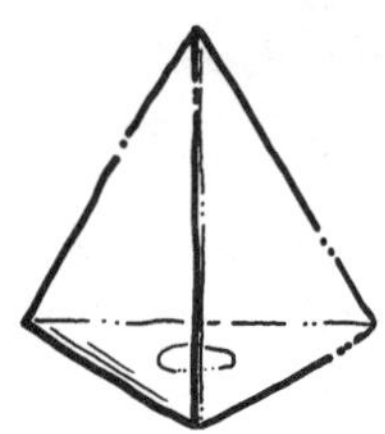
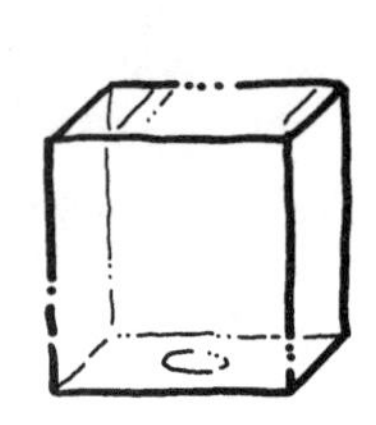

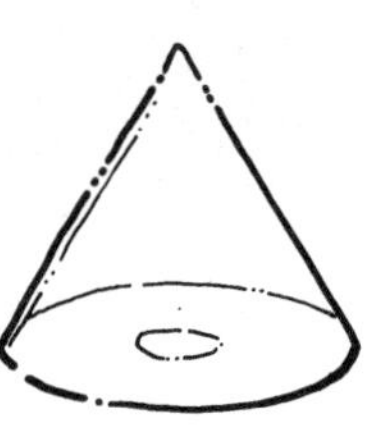

Filling Formulas

Solid	B Area of Base	× H height	= Calculated Volume (cm^3)	Measured Volume (cm^3 = mL)
Cube	l•w=B	5 cm		
Large Rectangular Prism	l•w=b	5 cm		
Small Rectangular Prism	l•w=B	5 cm		
Large Triangular Prism	(width • altitude)÷2=B	5 cm		
Small Triangular Prism	(width • altitude)÷2=B	5 cm		
Cylinder	πr^2=B	5 cm		
Hexagonal Prism	(edge)•(altitude)•3=B	5 cm		

Volumes of Cones and Pyramids

Solid	B Area of Base	× H height	÷	= Calculated Volume (cm^3)	Measured Volume (cm^3 = mL)
Square Pyramid	l•w=B	5 cm			
Triangular Pyramid	(width • altitude)÷2=B	5 cm			
Cone	πr =B	5 cm			

Topic
Geometric formulas

Key Question
How can you put three congruent pyramids together to form a cube?

Learning Goals
Students will:
- form a cube out of three congruent pyramids, and
- develop formulas based on a pyramid's relationship to the cube it occupies.

Guiding Documents
Project 2061 Benchmark
- *Calculate the circumferences and areas of rectangles, triangles, and circles, and the volumes of rectangular solids.*

*NCTM Standards 2000**
- *Create and critique inductive and deductive arguments concerning geometric ideas and relationships, such as congruence, similarity, and the Pythagorean relationship*
- *Develop strategies to determine the surface area and volume of selected prisms, pyramids, and cylinders*

Math
Geometry
measurement—volume
formulas of solids

Integrated Processes
Observing
Comparing and contrasting
Generalizing

Materials
Scissors
Glue

Background Information
A pyramid will have one-third the volume of a rectangular prism with the same size base and height. The general volume formula for a pyramid is: $V=\frac{1}{3}(b \bullet h)=\frac{1}{3}((l \bullet w) \bullet h)$.

The model used in this puzzle helps students understand this formula. The three congruent pyramids can be formed into a cube. The volume of the cube is the product of the length, width, and height. Since three congruent pyramids form this cube, each has a volume of a third of the cube.

Management
1. As a puzzle students will need time to explore solutions. It works well to have students construct the puzzle and then find solutions at home.
2. Use this as an individual puzzle with discussion of relationships after students have had time to construct the cube.

Procedure
1. Distribute puzzles, scissors, and glue. Have the students construct the puzzle according to the instructions on the page.
2. Encourage students to explore constructing a cube from the three pyramids.
3. Conduct a class discussion and have the class develop a formula to describe how they could use the volume of the cube to determine the volume of the pyramid.

Discussion
1. How did you construct a cube from the three pyramids?
2. Describe what shapes compose each face of the cube. [three sides squares, three sides two triangles]
3. How could you calculate the volume of the cube? [$l \bullet w \bullet h$]
4. How could you use the volume of the cube to determine the volume of the pyramid? [divide it by 3]
5. Write a formula to describe how you would find the volume of the pyramid from the volume of the cube. [$V=\frac{1}{3}(b \bullet h)=\frac{1}{3}((l \bullet w) \bullet h)=((l \bullet w) \bullet h)/3$]

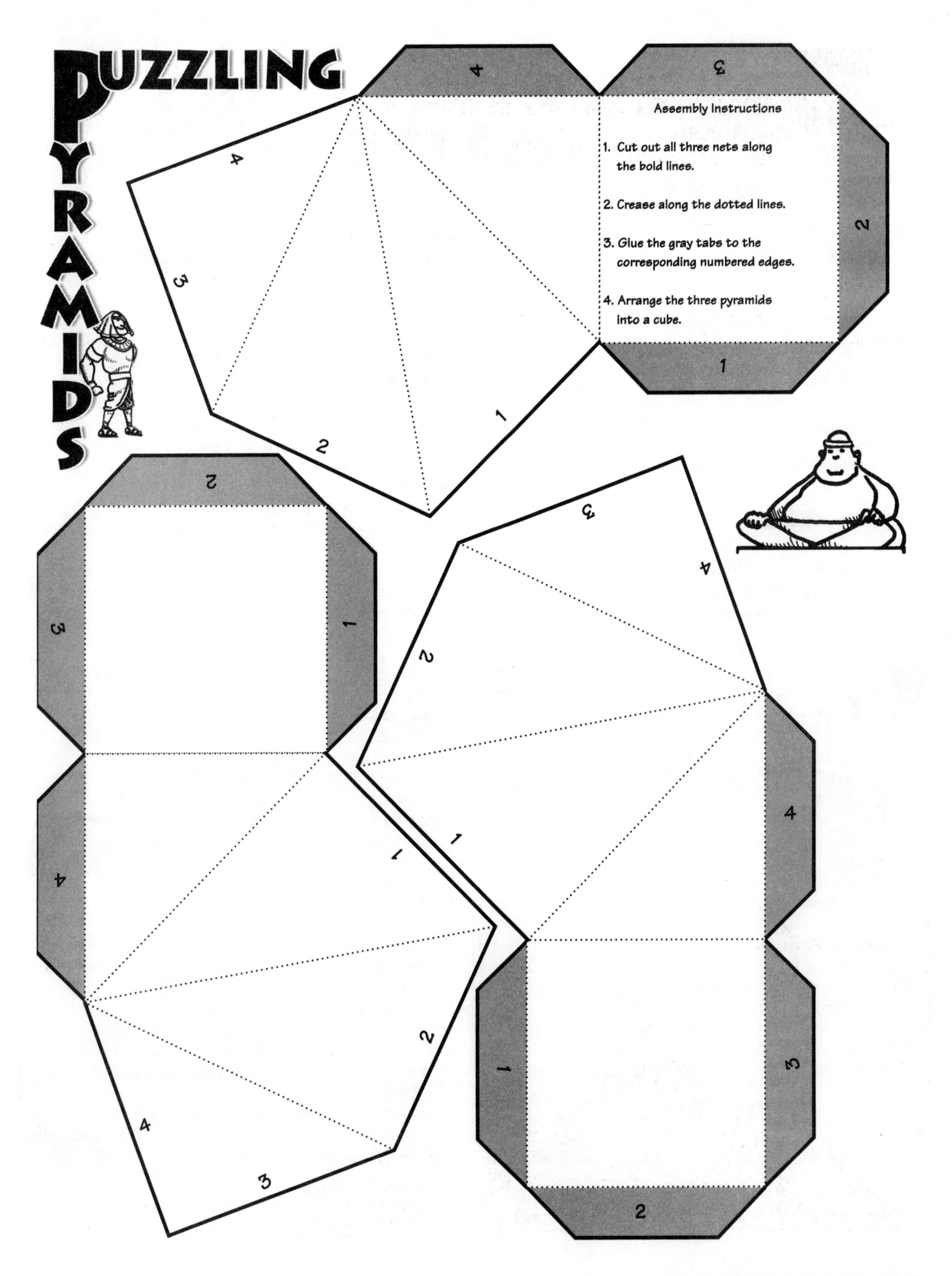
PUZZLING
PYRAMIDS
Assembly Instructions
1. Cut out all three nets along the bold lines.
2. Crease along the dotted lines.
3. Glue the gray tabs to the corresponding numbered edges.
4. Arrange the three pyramids into a cube.

Assembly Instructions

1. Cut out along the bold lines.
2. Crease along the dotted lines.
3. Glue the gray tabs to the edges to form a low box.
4. Arrange the three pyramids to make a cube with a base that fits in this box.

Assembly Instructions

1. Cut out along the bold lines.
2. Crease along the dotted lines.
3. Glue the gray tabs to the edges to form a low box.
4. Arrange the three pyramids to make a cube with a base that fits in this box.

Assembly Instructions

1. Cut out along the bold lines.
2. Crease along the dotted lines.
3. Glue the gray tabs to the edges to form a low box.
4. Arrange the three pyramids to make a cube with a base that fits in this box.

Assembly Instructions

1. Cut out along the bold lines.
2. Crease along the dotted lines.
3. Glue the gray tabs to the edges to form a low box.
4. Arrange the three pyramids to make a cube with a base that fits in this box.

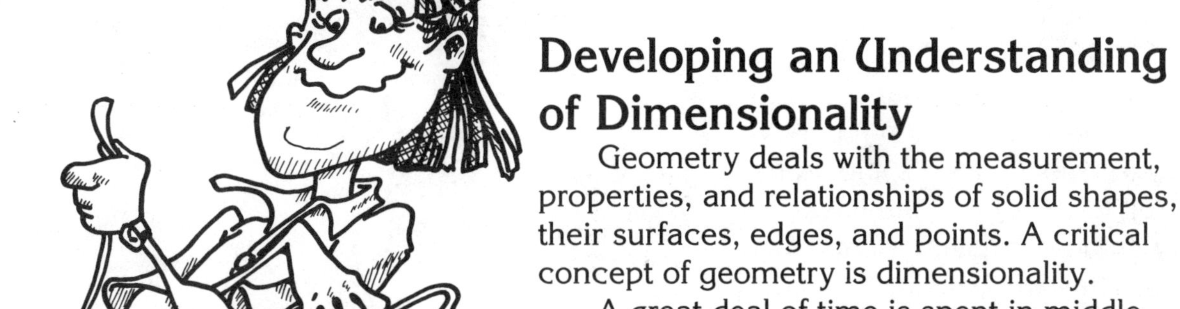

Developing an Understanding of Dimensionality

Geometry deals with the measurement, properties, and relationships of solid shapes, their surfaces, edges, and points. A critical concept of geometry is dimensionality.

A great deal of time is spent in middle school mathematics calculating dimensional measurements by using formulas. Typical problems show sketches of rectangles and boxes and ask students to calculate the perimeters, areas, or volumes. Unless students have had extensive experiences with real objects, they will not recognize the dimensional differences of perimeter, area, and volume. In this situation students will select a formula they remember, hoping it is the right choice. The students do the calculation correctly but often choose the wrong formula. They need an understanding of dimensionality before they begin to deal with the numbers related to measurement.

Understanding dimensionality would mean recognizing that the length of a ribbon is one dimensional even though it would stretch across all six faces of the box. A student reporting the length and width of a sheet of wrapping paper required to cover the box demonstrates the two-dimensional nature of surface area. Checking that a gift is shorter, narrower, and less deep than the box shows the three-dimensional nature of volume.

Physical Manipulatives

Geometry by nature deals with the physical world. If the measures of geometry are to have meaning, they must be experienced with physical manipulatives as they are connected to numbers. The physical objects used must lend themselves to numeration to provide a natural tie for students.

The nature and measurements of three-dimensional objects can be developed by using cubes. The mosaic of squares on the surface provide a natural quantitative measure of area needed to cover the solid. By cutting grid paper to cover the surface, the two-dimensional nature is emphasized. The space-filling nature of the cubes provides a meaning to volume. The description of the filled solid by length, width, and height provides more meaning to the concept of three dimensional.

Abstract Formulas

Students' initial work with the measurement of geometric figures should be done at the manipulative level where the dimensional measurement can be seen as well as counted. As students' experiences grow, they will be able to work at the representational level because they can visualize what is not shown in the picture. Through repetitive experiences, students should be asked to determine the linear, area, and volume measurements of figures. They will develop procedures for finding these measurements efficiently. Formulas should be developed by students discussing their procedures and then translating them into math sentences.

To determine perimeters, many students count the length of each side and add them together. For a rectangle this is recorded L + W + L + W = P. Other students will say there are two lengths and widths, so they double them and add them together. This procedure is recorded as (2 x L) + (2 x W) = P. Some students will say they add the distances halfway around and then double the sum to get the total resulting in a procedure written 2(L + W) = P. Geometric formulas made from students' procedures are used successfully because they are tied to meaningful experiences that can be revisited.

Experienced students looking at a representation of a 5 x 4 x 3 unit rectangular prism will think, "On the front face there are five squares in a row and three rows like that. That's 15 squares on the front." They recognize the two-dimensional nature of surface area. They go on, "The back is like the front, so I just double the 15. If I follow the same procedure for the other four faces and add them together, I know how many squares are needed to cover the box." Because of experience, the students understand that there is a back even though it cannot be seen. When they are shown that the procedure could be symbolized in the following way, 2(L x W) + 2(L x H) + 2(W x H) = SA, it makes sense; experience has made the formula meaningful. Conversely, the formula is meaningless without the experience.

Background Article

Boxes, Boxes, Boxes
Part One

Packing, volume, and surface area problems arise in the real-world context of a shipping department. The shippers are responsible for receiving and shipping a variety of materials. Nearly all of these items come in bulk quantities. Some of them need to be repackaged in smaller quantities, others are assembled into kits, and still other items are simply shelved. Eventually, all of these items will find their way into one of the many orders that are shipped throughout the country.

Were you to drop by and observe the shippers at work, you would soon conclude that their main activity is *packing*. Different orders require boxes of different sizes. The best box for a given order can be selected from a variety of sizes available in neatly arranged compartments. For example, if an order calls only for books, a box having an $11\frac{1}{4}$" by $8\frac{3}{4}$" base will be selected. The heights of these boxes vary from one inch to 12 inches. The one-inch box will hold two or three books, while the 12-inch box will hold anywhere from 24 to 40 books.

New items are always being added to the line of products. This often requires the shipping department to design special boxes in which to pack these items. For example, wooden cubes were added to the product line. These cubes came from the supplier in large boxes containing five or six thousand cubes. Since customers ordering cubes don't want to buy that many at a time, the shipping department repackages the cubes in bags of 500. If a shipment calls for just one of these bags of cubes, there is a 12" x 10" x 6" box into which the bag fits very nicely. If a shipment calls for two bags, they are packed in a 12" x 12" x 10" box.

The following describes the solution of a packing problem that could very well occur within the context of a shipping department. Here's the scenario:

Twelve-Cube Packing Problem

In response to customer demand, a box is needed that will hold 12 one-inch cubes required for a puzzle. Anticipating that there will be many requests for the cubes, two of the shippers have been asked to design a small box to hold the 12 cubes. Since they plan to shrink-wrap the boxes, the box design does not need a lid, but the design should utilize the least amount of material.

As the shippers began to think about the problem, they quickly realized that what they needed to do was simply design a box with a volume of 12 cubic inches—the volume being equal to the number of cubes. They decided to get a set of the 12 cubes and actually stack them in different ways to help them determine what the possibilities for different boxes might be. They found basically four different arrangements of the 12 cubes: 1 x 1 x 12, 1 x 2 x 6, 1 x 3 x 4, and 2 x 2 x 3.

Since this was their first design project, the shippers decided to actually build each of the possible boxes. They found some sheets of one-inch graph paper and began to lay out the box having dimensions 2 x 2 x 3. Their layout looked something like the one in *Figure 1*. With this arrangement, they were able to construct the box by having folds at four of the edges, leaving the other four edges to be taped. Since one of the six faces of the box is to be open, their layout was done in such a way that the bottom was in the center of the layout, with the other four faces attached to it.

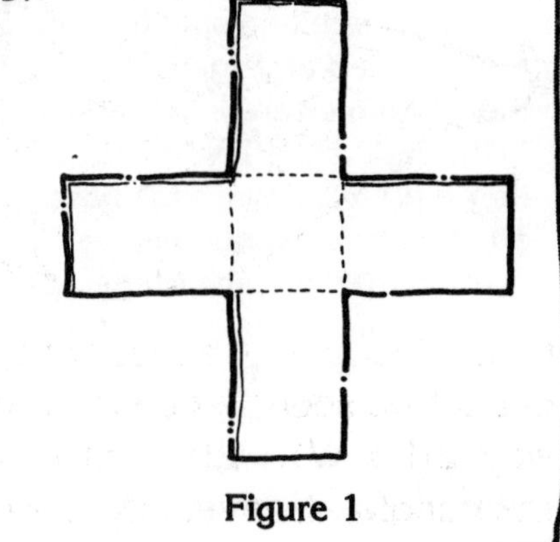

Figure 1

Their first box had a 2 x 2 bottom, 2 x 3 sides, and a 2 x 2 opening. They knew from stacking the cubes that they could also construct a box so that the bottom is 2 x 3, the sides 2 x 2, and the opening 2 x 3 (*Figure 2).*

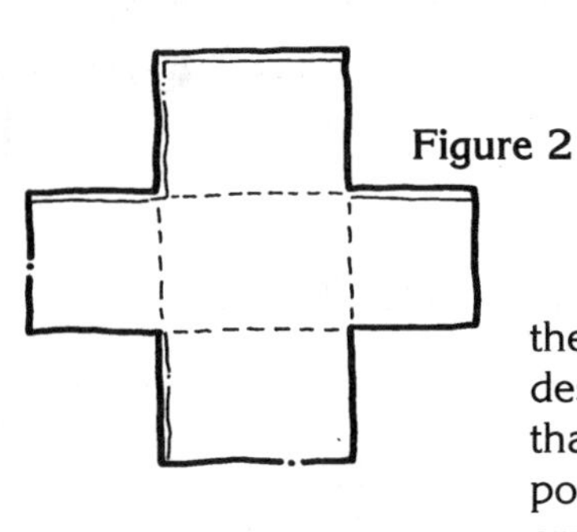

Figure 2

Figure 3

In the process of constructing the two boxes (*Figure 3),* the designers assured themselves that these were the only two possible boxes for a 2 x 2 x 3 arrangement of the cubes. While the boxes have the same basic shape, they differ in the size of the open face and in the orientation of the box. They reasoned that no other open boxes would be possible since there are only two kinds of faces: 2 x 2 and 3 x 2.

The designers next tackled the construction of 1 x 1 x 12 boxes. Their first thought was to lay these out in the same way they had done for the 2 x 2 x 3. In other words, to first lay out the box with a 1 x 1 bottom and 1 x 1 open top and then lay out a second one with a 1 x 12 bottom and 1 x 12 open top. Before doing the layout on the one-inch graph paper, one shipper decided to quickly sketch what it might look like (*Figure 4).* After seeing how much paper it would take to lay them out in this way, the designers did some experimenting with other possible ways to lay out the sides and still only need to tape four edges. They discovered a layout that would save a lot of paper (*Figure 5).*

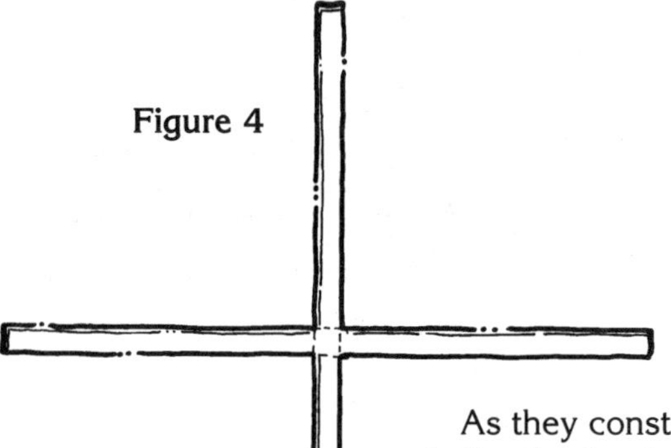

Figure 4

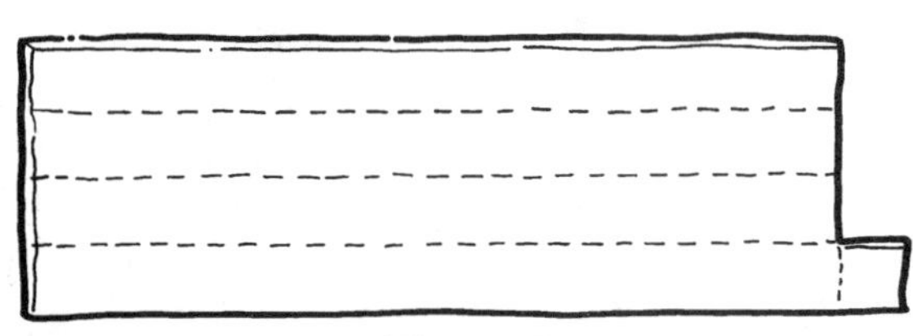

Figure 5

As they constructed these two boxes, the two shippers talked about how similar the 1 x 1 x 12 problem is to the 2 x 2 x 3 problem. They noticed that in each case, the base of the box is the same size and shape as the open top. They also noticed that when the base and top are squares, the other four sides are all four rectangles of the same size and shape. For the 2 x 2 x 3 arrangement, the rectangles are 2 x 3, and in the case of the 1 x 1 x 12, they are 1 x 12. When the base and top are not squares, they noticed that for each of the two arrangements, two of the other four sides are squares and two are rectangles having the same size and shape as the bottom and top.

Still remaining to be laid out and constructed were boxes for the 1 x 2 x 6 and 1 x 3 x 4 arrangements of the cubes. As the shippers examined cubes stacked in these two arrangements, they noticed that in each case, there are no square faces. Moreover, the faces are made up of three different rectangles. In the case of the 1 x 2 x 6 arrangement, the rectangular faces are 1 x 2, 1 x 6, and 2 x 6; and in the case of the 1 x 3 x 4, the faces are 1 x 3, 1 x 4, and 3 x 4. Finally, they noticed that for each of these arrangements there are three different possibilities for open faces, and consequently three different boxes can be constructed for each. *Figure 6* and *Figure 7* show the box layouts for the two arrangements. *Figures 8* and *9* show the finished boxes.

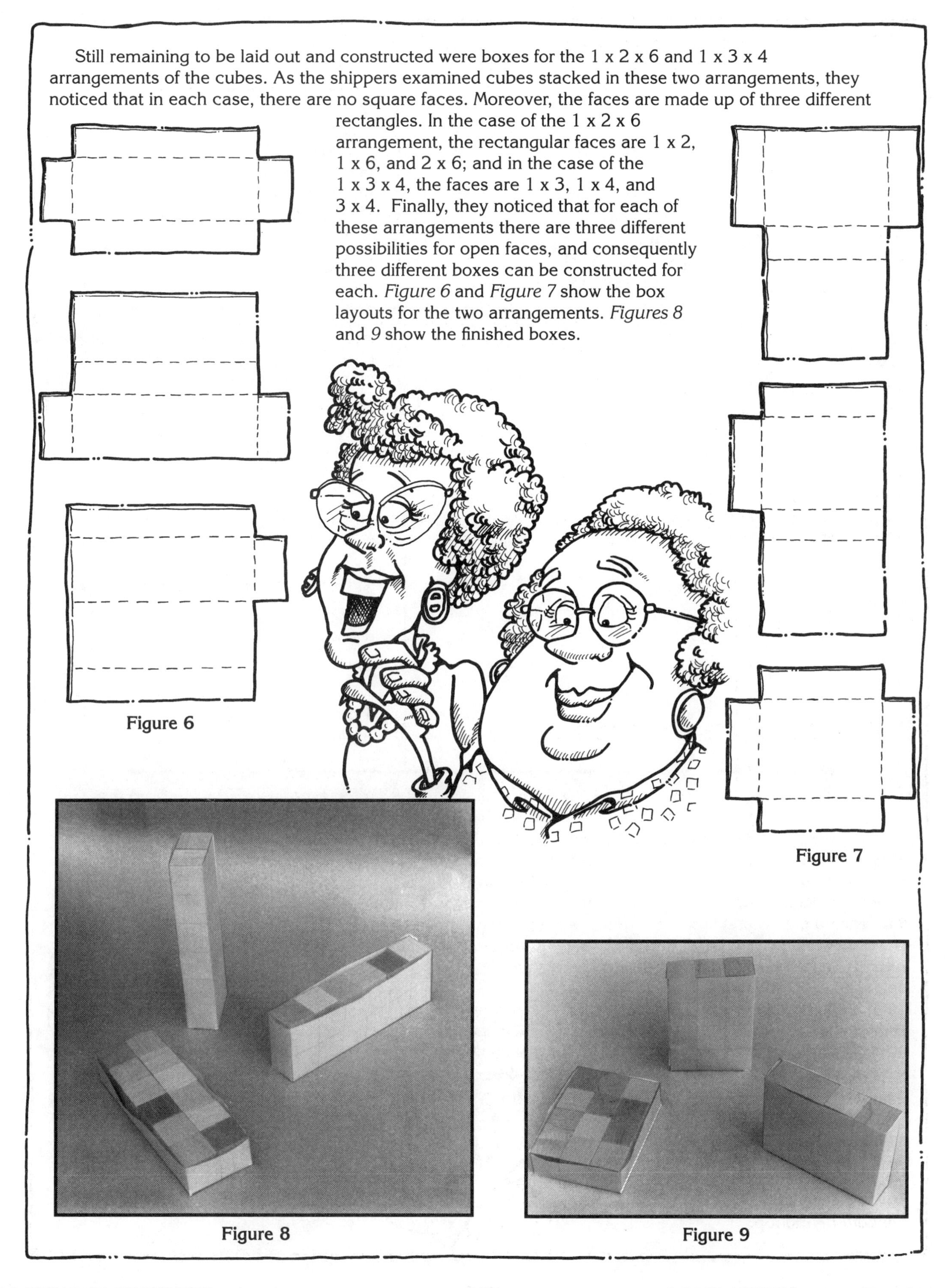

Figure 6

Figure 7

Figure 8

Figure 9

Having now constructed all of the ten possible boxes, the time had come to make a decision about which should be the one used to pack and ship the 12 cubes. They remembered that the decision was to be based on which box requires the least amount of material to construct. With this in mind, the designers created the following table to help them easily view all the possibilities.

Arrangement of Cubes	Bottom/Top Face	Total Surface Area
1 x 1 x 12	1 x 1	1(1 x 1) + 4(1 x 12) = 49
1 x 1 x 12	1 x 12	2(1 x 1) + 3(1 x 12) = 38
2 x 2 x 3	2 x 2	1(2 x 2) + 4(2 x 3) = 28
2 x 2 x 3	2 x 3	2(2 x 2) + 3(2 x 3) = 26
1 x 2 x 6	1 x 2	1(1 x 2) + 2(1 x 6) + 2(2 x 6) = 38
1 x 2 x 6	1 x 6	1(1 x 6) + 2(1 x 2) + 2(2 x 6) = 34
1 x 2 x 6	2 x 6	1(2 x 6) + 2(1 x 2) + 2(1 x 6) = 28
1 x 3 x 4	1 x 3	1(1 x 3) + 2(1 x 4) + 2(3 x 4) = 35
1 x 3 x 4	1 x 4	1(1 x 4) + 2(1 x 3) + 2(3 x 4) = 34
1 x 3 x 4	3 x 4	1(3 x 4) + 2(1 x 3) + 2(1 x 4) = 26

They were surprised at how great a variation in surface area there is among boxes—from 26 to 49. Before computing the various areas, they had expected that boxes having the larger openings would have the least surface areas. They were surprised to note that this was not necessarily so. For example, one of the 1 x 1 x 12 boxes has a 1 x 12 opening and a surface area of 38, while one of the 2 x 2 x 3 boxes with a 2 x 2 opening has a surface area of 28.

The information in the chart along with the boxes they had constructed gave our designers the information they needed to make a recommendation. Two of the boxes met the "least surface area" criteria: the 1 x 3 x 4 box with a 3 x 4 opening and the 2 x 2 x 3 box with a 2 x 3 opening. The shippers decided to recommend the 2 x 2 x 3 box simply because it had the smaller opening.

Topic
Geometry

Key Question
What are all the possible nets with only square faces that can be designed for a cube?

Learning Goals
Students will:
- cut out a variety of nets that can be used to wrap a cube, and
- organize the nets to determine if all the possible nets have been found.

Guiding Documents
Project 2061 Benchmark
- *Some shapes have special properties: Triangular shapes tend to make structures rigid, and round shapes give the least possible boundary for a given amount of interior area. Shapes can match exactly or have the same shape in different sizes.*

*NCTM Standards 2000**
- *Understand relationships among the angles, side lengths, perimeters, areas, and volumes of similar objects*
- *Use two-dimensional representations of three-dimensional objects to visualize and solve problems such as those involving surface area and volume*

Math
Geometry
  surface area
  nets

Integrated Processes
Observing
Collecting and organizing data
Interpreting data
Drawing conclusions

Materials
Hex-a-link or 2-cm cubes
Scissors

Background Information
A net is the two-dimensional form that can be folded to surround a three-dimensional space. A cube is a three-dimensional figure made of six square faces. The net for a cube is six squares where each square shares a common edge with another square. Below are all eleven possible nets for a cube made of squares.

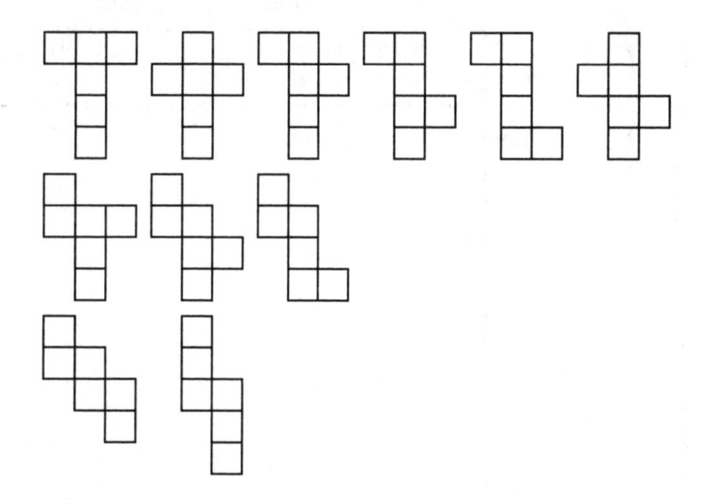

Management
1. This activity may be done individually by students outside of class as written. If time is short, the teacher may choose to have students work cooperatively and share solutions as they are discovered.
2. Make sure students understand that each square *must* share an edge with another square. Students often try to place the squares diagonally across from each other and then cut them so that the squares remain attached.
3. Students will need experiences of folding nets around cubes to develop spatial abilities. When they have folded a net, encourage them to label the six faces (bottom, top, back, front, right, left) will help them begin to fold the nets in their minds. As students progress, many with the help of labeling, will be able to determine if a net works without actually folding it.

Procedure
1. Provide each student with a cube, scissors, and grid paper.
2. Ask the *Key Question:* What are all the possible nets with only square faces that can be designed for a cube? Have the students predict the number of nets possible in the situation.
3. Allow students time to explore the question using their materials to build trial nets. If students are

floundering on the instructions, model a solution such as the "T" or the "cross." To provide sufficient time, encourage the students to work at finding as many as they can at home.

4. When students have a number of solutions, ask them to share the solutions with the class. Each student should build the net and confirm that it works. Students can display the solutions as they are found to encourage progress in finding them all.
5. As students conclude that they have found all the solutions, have them order and arrange them by similarities or patterns to see that they have found all eleven. Many students will sort them like the key in *Background Information* by "four-in-a-row," "three-in-a-row," and "stairs." By making such arrangements, students can spot patterns that allow them to see missing solutions.
6. When the class has found all the solutions, discuss the patterns they have found.

Discussion

1. How many squares did it take to make each of the nets? [6]
2. What patterns did you discover to help you find new solutions or verify that you had them all?
3. If each net were cut from a rectangle of cardboard to make a box, how big would the cardboard need to be? [12 squares, 3 x 4; 10 squares, 2 x 5]

Extension

Have students consider the situation where a company wants to cut nets for a cubic box out of sheets of cardboard the size of a piece of paper. Many nets will fit on each sheet. The nets can be nested so there is less waste. Have the students explore and report which single net will be the most efficient by having the least waste.

* Reprinted with permission from *Principles and Standards for School Mathematics,* 2000 by the National Council of Teachers of Mathematics. All rights reserved.

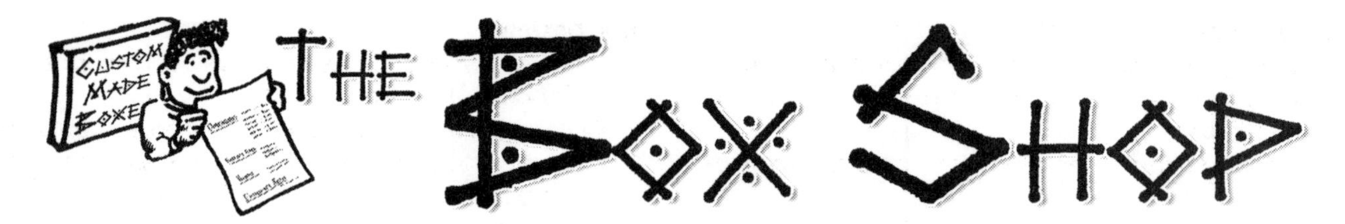

Topic
Geometry

Key Question
If you were making a box for 24 cubes, what is the least number of squares of cardboard it would take to cover them?

Learning Goals
Students will:
- determine surface area and volume with a situation where the volume is constrained but the surface area varies, and
- describe their processes in equation form.

Guiding Documents
Project 2061 Benchmark
- *Some shapes have special properties: Triangular shapes tend to make structures rigid, and round shapes give the least possible boundary for a given amount of interior area. Shapes can match exactly or have the same shape in different sizes.*

*NCTM Standards 2000**
- *Understand relationships among the angles, side lengths, perimeters, areas, and volumes of similar objects*
- *Use two-dimensional representations of three-dimensional objects to visualize and solve problems such as those involving surface area and volume*

Math
Geometry
 surface area
 volume
 efficiency

Integrated Processes
Observing
Comparing and contrasting
Collecting and organizing data
Interpreting data
Drawing conclusions

Materials
Hex-a-link or 2 cm cubes

Background Information

The efficiency of a container is based on the ratio of surface area to volume. The container with the smallest ratio uses the least surface area to cover a given volume. A sphere is the most efficient solid. A cube is the most efficient rectangular solid. The closer a rectangular solid is to a cube, the more efficient it is.

There are six ways to form a rectangular solid from 24 cubes. The one that is closest to cube shape is the most efficient.

Dimensions	Surface Area	Volume	Efficiency Ratio (SA/V)
1 x 1 x 24	98 sq. units	24 cu. units	4.0833...
1 x 2 x 12	76 sq. units	24 cu. units	3.166...
1 x 3 x 8	70 sq. units	24 cu. units	2.9166...
1 x 4 x 6	68 sq. units	24 cu. units	2.833...
2 x 2 x 6	56 sq. units	24 cu. units	2.33...
2 x 3 x 4	52 sq. units	24 cu. units	2.166...

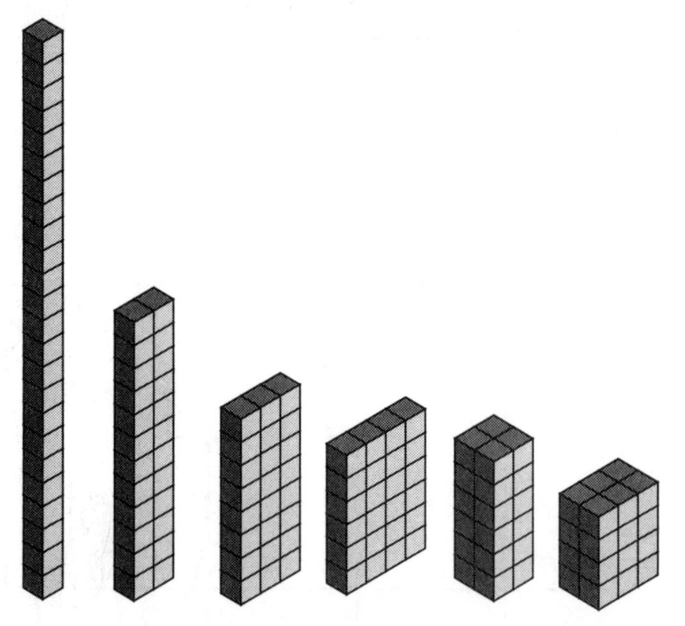

After students have had some experience determining surface area by counting, it is important to have them summarize methods to shorten the process. These can be translated into equations that will have meaning to the students. Many students will talk about counting the top and then doubling it to get the bottom. They will follow a similar process on the front/back and sides. Their equation translates to: $SA=2(l \bullet w)+2(w \bullet h)+2(l \bullet h)$. Other students will determine the sum of the areas of the top, front, and side. They double this sum because it only includes half of the faces. They get a different but equivalent equation: $SA=2(lw+wh+lh)$.

Management

1. Each group of students will need 24 cubes. Groups of two are ideal, but groups of four are satisfactory.
2. Students should be familiar with isometric drawings before doing this activity.

Procedure

1. After introducing the investigation with the *Key Question*, have the students construct each of the six possible rectangular solids.
2. Referring to the model they have constructed, ask them to make an isometric drawing.
3. Using the model of the rectangular solid, tell the students to count and record the dimensions, surface area, and volume.
4. Have the class discuss how they can be sure they have all the possible solutions.
5. As a class, discuss short cuts used in counting surface area and how they could be translated into equations.
6. Have the students compare the surface areas and drawings of the solids and discuss how one could tell by looking at a set of solids that have the same volume which will have the least surface area.
7. Ask the students to calculate the efficiency ratio of each solid and discuss what the ratio is telling them about each box.

Discussion

1. How can you be sure you have all the possible solutions? [organize the list]
2. What are some short cuts you used to count the surface area of the cubes? (See *Background Information.)*
3. How could you describe the process you used to count the surface area of a box with an equation? (See *Background Information.)*
4. How is the shape of a solid related to its surface area? [The more cubic the shape, the less surface area it has.]
5. How can you tell which solid makes the most use of the squares that cover it? [least surface area]
6. What does the ratio of the surface area to volume tell you about the shape? [how many squares on the surface are needed for each cube in the shape]

Extension

Have students consider this situation:

Your company wants to waste as little cardboard as possible. It is willing to cut more than one net from a rectangular piece of cardboard. Design a layout of nets for 24 cubes that would waste the least cardboard. All the nets must be for the same box, but do not need to be the same net.

THE BOX SHOP

Dimensions Length x Width x Height	Surface Area (SA)	Volume (V)	Efficiency Ratio SA/V = Decimal Equivalent
X X			=
X X			=
X X			=
X X			=
X X			=
X X			=

Make isometric drawings of the rectangular solids you made.

Topic
Geometry

Focused Task
Your company manufactures cubic crystals measuring a foot on each side. The protective shipping container for the crystals costs a dollar per square foot to manufacture. The shipping company will only accept packages with eight feet in total linear dimensions (l+w+h). You have been asked to design the container that will provide the lowest cost per crystal cube. Prepare a report (presentation) showing all the possibilities and the reasoning for your recommendation.

Learning Goal
Students will determine the most efficient box for cubes with a constraint of total linear dimensions.

Guiding Documents
Project 2061 Benchmark
- *Some shapes have special properties: Triangular shapes tend to make structures rigid, and round shapes give the least possible boundary for a given amount of interior area. Shapes can match exactly or have the same shape in different sizes.*

*NCTM Standards 2000**
- *Understand relationships among the angles, side lengths, perimeters, areas, and volumes of similar objects*
- *Use two-dimensional representations of three-dimensional objects to visualize and solve problems such as those involving surface area and volume*

Math
Geometry
  surface area
  volume
  efficiency

Integrated Processes
Observing
Comparing and contrasting
Collecting and organizing data
Interpreting data
Drawing conclusions

Materials
Hex-a-link or 2 cm cubes

Background Information
A constraint of total linear length is often put on items shipped by airlines. It provides a context where both volume and surface area are calculated. The solutions for this investigation are listed below.

Dimensions	Surface Area	Volume	Efficiency Ratio (SA/V)
1 x 1 x 6	26 sq. units	6 cu. units	4. 33...
1 x 2 x 5	34 sq. units	10 cu. units	3.4
1 x 3 x 4	38 sq. units	12 cu. units	3.166...
2 x 2 x 4	40 sq. units	16 cu. units	2.5
2 x 3 x 3	42 sq. units	18 cu. units	2.33...

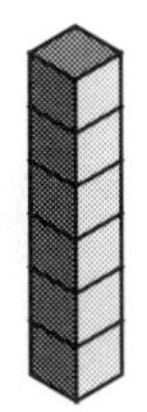
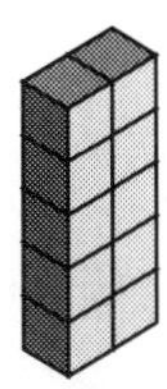
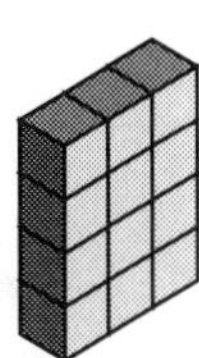

Management
1. Each group of students will need 18 cubes.
2. This activity can simply be done as a reinforcement investigation as it is written. Some teachers may choose to treat it as an open-ended project or performance task for an assessment. In these cases, introduce the *Focused Task*, provide access to the material, and allow students to work. When done, have students present their work and conclusions as a written report or oral presentation.

Procedure
1. After introducing the investigation with the *Focused Task*, have the students construct each of the five possible rectangular solids.
2. Referring to the model they have constructed, ask students to make an isometric drawing, calculate and record the dimensions, surface area, and volume.
3. Have the class discuss how they can be sure they have all the possible solutions.
4. Ask the students to calculate the efficiency ratio of each solid and discuss what the ratio is telling them about each box.
5. Have the students compare the efficiency ratios and drawings of the solids, and discuss how they can tell by looking at a set of drawings which would be most efficient.

Discussion

1. How can you be sure you have all the possible solutions? [Make an organized list.]
2. How is the shape of a solid related to its efficiency ratio? [The more cubic the shape, the more efficient it is and the lower its efficiency ratio is.]
3. What does the efficiency ratio mean in dollars to your company? [dollar cost for each crystal (unit) in the container]
4. Explain what dimension container you would recommend to your supervisor and explain how you came to this decision. [2 x 3 x 3, lowest efficiency ratio = lowest cost per unit]
5. How could you make similar decisions in the future without all the calculations? [choose the most cubic container]
6. The airlines restrict luggage by total linear length. Explain what shape luggage you would choose to carry the most volume of clothes on your next trip. [cubic]
7. Why might the most efficient luggage not be the most practical? [awkward to carry]

Extension

Have students design the most efficient net for the most efficient container. Before they fold up their net, have them put labeling on the four sides so that when the net is folded, the writing is right side up.

* Reprinted with permission from *Principles and Standards for School Mathematics,* 2000 by the National Council of Teachers of Mathematics. All rights reserved.

Make isometric drawings of the rectangular solids you made.

Dimensions Length x Width x Height	Surface Area (SA)	Volume (V)	Efficiency Ratio SA/V = Decimal Equivalent
X X			=
X X			=
X X			=
X X			=
X X			=
X X			=

Topic
Geometry

Focused Task
A company is going to ship packages of 12 one-inch cubes required for a puzzle. Working in the shipping department you have been asked to recommend a design for a small box to hold the 12 cubes. Since the box will be shrink wrapped, it does not need a lid and only 5 faces will be covered. You need to find the design that will utilize the least amount of material.

Learning Goal
Students will determine the most efficient open-faced box for 12 cubes.

Guiding Documents
Project 2061 Benchmark

- *Some shapes have special properties: Triangular shapes tend to make structures rigid, and round shapes give the least possible boundary for a given amount of interior area. Shapes can match exactly or have the same shape in different sizes.*

*NCTM Standards 2000**

- *Understand relationships among the angles, side lengths, perimeters, areas, and volumes of similar objects*
- *Use two-dimensional representations of three-dimensional objects to visualize and solve problems such as those involving surface area and volume*

Math
Geometry
 surface area
 volume
 efficiency

Integrated Processes
Observing
Comparing and contrasting
Collecting and organizing data
Interpreting data
Drawing conclusions

Materials
Hex-a-link or 2 cm cubes

Background Information
Surface area in math class means all six faces of a rectangular solid. In many situations, the surface area under consideration is not the total surface area of an object. This investigation looks into the effects of a missing face on surface area.

Arrangement of 12 Blocks	Open Face of Box Length X Width = Area	Surface Area of 5-Sided Box
Dimensions 1 X 1 X 12 Surface Area 50 sq. in.	1 X 1 = 1	49 sq. in.
	1 X 12 = 12	38 sq. in.
	X =	
Dimensions 1 X 2 X 6 Surface Area 40 sq. in.	1 X 2 = 2	38 sq. in.
	1 X 6 = 6	34 sq. in.
	2 X 6 = 12	28 sq. in.
Dimensions 1 X 3 X 4 Surface Area 38 sq. in.	1 X 3 = 3	35 sq. in.
	1 X 4 = 4	34 sq. in.
	3 X 4 = 12	26 sq. in.
Dimensions 2 X 2 X 3 Surface Area 32 sq. in.	2 X 2 = 4	28 sq. in.
	2 X 3 = 6	26 sq. in.
	X =	

In this case, there are two equally efficient arrangement with a 5-sided surface area of 26 square inches. The final decision may be on which net is most cost effective.

Management

1. Each group of students will need 12 cubes. The students should be made aware that the cubes are models and not the actual size of cubes in the scenario.
2. This activity can simply be done as a reinforcement investigation as it is written. Some teachers may choose to treat it as an open-ended project or performance task for an assessment. In these cases, introduce the *Focused Task*, provide access to the material, and allow students to

work. When done, students can present their work and conclusions as a written report or oral presentation.

Procedure

1. After introducing the investigation with the *Focused Task*, have the students construct each of the four possible rectangular solids.
2. Referring to the model they have constructed, ask the students to make an isometric drawing, calculate and record the dimensions and surface areas.
3. Tell the students to consider each size face on the model, record its dimensions and area, and determine the surface area of the box if that face were left open.
4. Have the students compare the surface areas of the five-sided boxes and drawings of the solids and explain the variety of surface areas.

Discussion

1. What are all the possible arrangements for boxes made of 12 cubes?
2. What are all the different possible surface areas for the topless boxes?
3. Why are there different surface areas of the same box arrangements?
4. What are some reasons boxes may not always be made most efficiently? [shape of what is in it, convenience, appearance]

Extension

There are two different arrangements that are equally efficient. Have the students investigate for which arrangement it would be possible to construct the most efficient net.

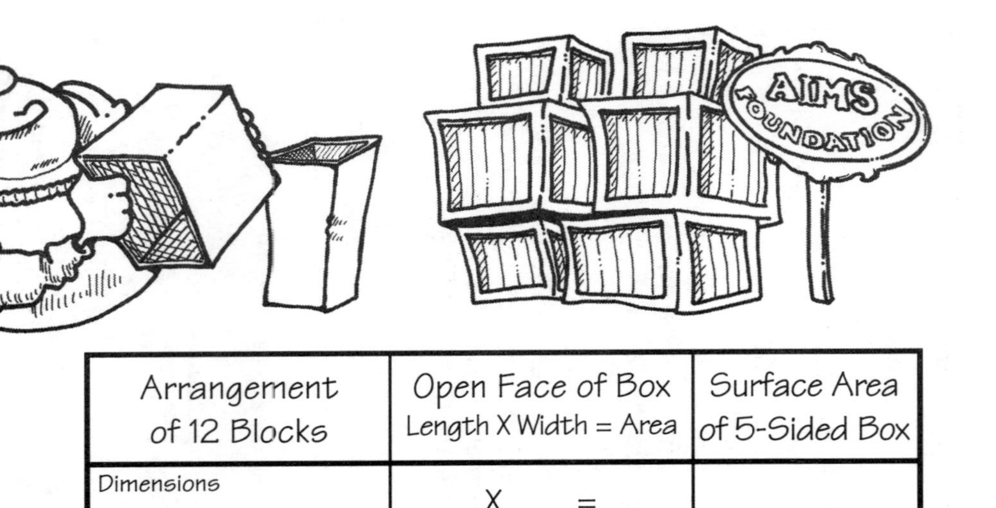

Make isometric drawings of the rectangular solids you made.

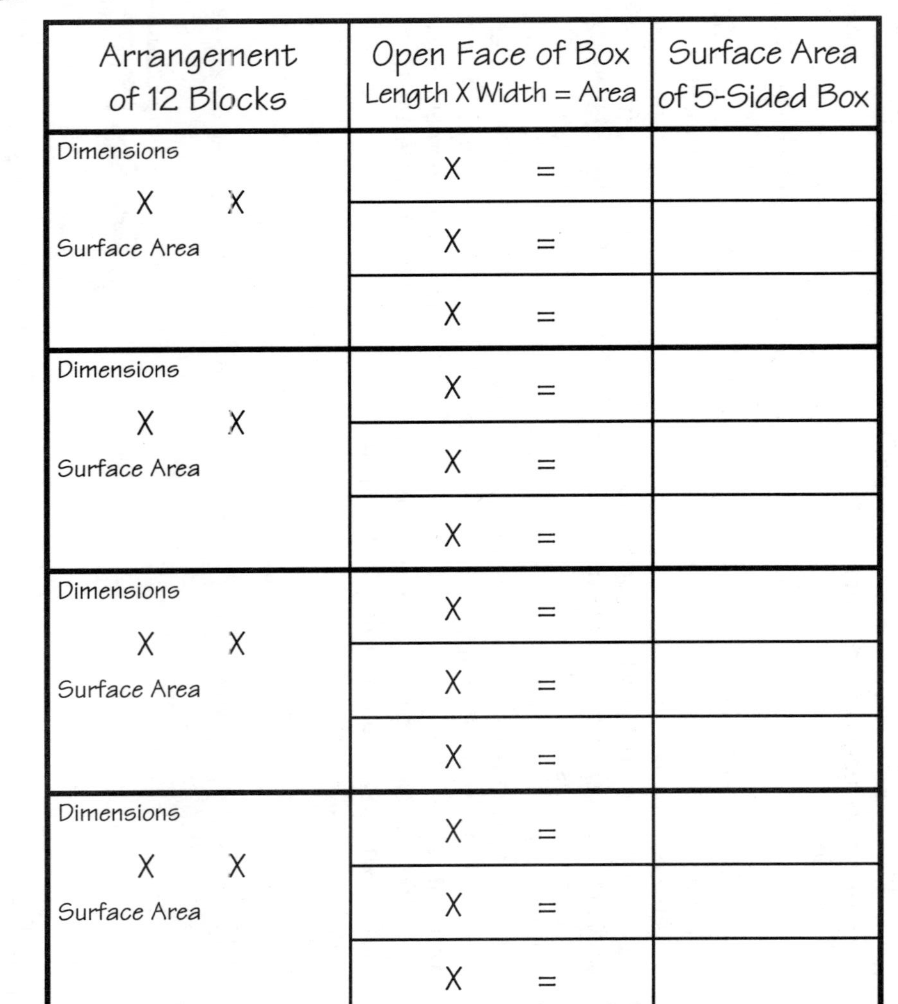

Arrangement of 12 Blocks	Open Face of Box Length X Width = Area	Surface Area of 5-Sided Box
Dimensions X X Surface Area	X =	
	X =	
	X =	
Dimensions X X Surface Area	X =	
	X =	
	X =	
Dimensions X X Surface Area	X =	
	X =	
	X =	
Dimensions X X Surface Area	X =	
	X =	
	X =	

Background Article

Boxes, Boxes, Boxes
Part Two

Part One of *Boxes, Boxes, Boxes* described the work of two shippers as they designed an open box to hold 12 individual cubes. They were asked to find a design that utilizes the least amount of material in its construction. They found that there are ten different boxes, each of which will hold the 12 cubes. The amount of surface area required to construct these boxes varies from a minimum of 26 square inches to a maximum of 49 square inches. It turns out that there are two different boxes with the minimum surface area of 26 square inches: the 2 x 2 x 3 box with a 2 x 3 opening, and the 1 x 3 x 4 box with a 3 x 4 opening. The box with dimensions 1 x 1 x 12 and a 1 x 1 open face had the maximum surface area.

With the success of their first box design, these two shippers were prepared when they were presented with a six-piece mechanical puzzle (*Figure 1*). The puzzle pieces were constructed by gluing cubes together. It took 36 cubes to construct all six pieces — two of the pieces required five cubes, two required six, and two required seven. The challenge of this puzzle is to put the six pieces together to form a rectangular solid arrangement. Without showing them how to put the pieces together to form such an arrangement, they were asked to design a box that would hold the new puzzle. However, the box needed to meet the same requirements as was the case with the box for the 12 cubes. In other words, it should be an open box (no lid) with a volume equal to the number of cubes required to construct the pieces, and it should be constructed in a manner requiring the least amount of material.

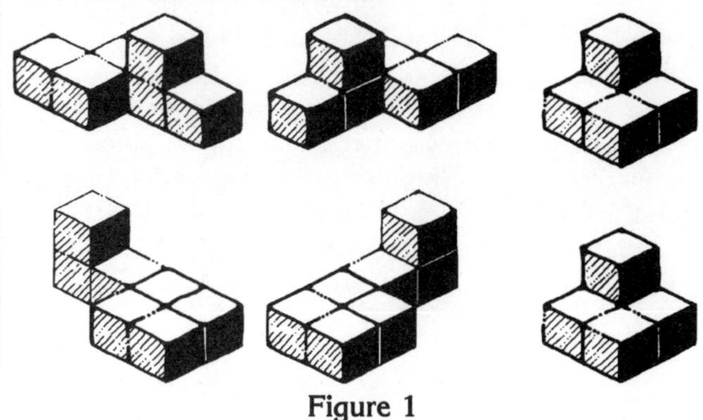

Figure 1

After several frustrating attempts to put the pieces together, the two shippers decided to proceed as they had with the 12-cube box. They first looked at all the different boxes that could be constructed to hold 36 cubes. Finding a bag of individual cubes, they started to put them together in rectangular solid arrangements *(Figure 2)*.

Figure 2

They found eight different rectangular solid arrangements for the 36 cubes:

1x1x36	1x6x6
1x2x18	2x2x9
1x3x12	2x3x6
1x4x9	3x3x4

Table 1

The shippers quickly realized that given the puzzle pieces, it would be impossible to pack them into a box having a depth, width, or length of one. This eliminated the first five arrangements on their list from further consideration. As they had done with the 12-cube box, they decided as a next step to construct paper boxes for the remaining three arrangements. *(Table 2* and *Figure 3* show their work in designing and constructing the boxes.)

arrangement	open face
2x2x9	2x2
2x2x 9	2x 9
2x3 x 6	2x3
2x3 x 6	3x6
2x3x 6	2x6
3x3x 4	3x 3
3x3x 4	3x 4

Table 2

Figure 3

With the boxes in view, it was easy for them to determine which box or boxes would require the least amount of material to construct. It turned out that two different boxes, each with a surface area of 54, satisfied this requirement: the 2 x 3 x 6 box with a 3 x 6 opening and the 3 x 3 x 4 box with a 3 x 4 opening. Their actual calculations to find the surface area of each box are shown in *Table 3.*

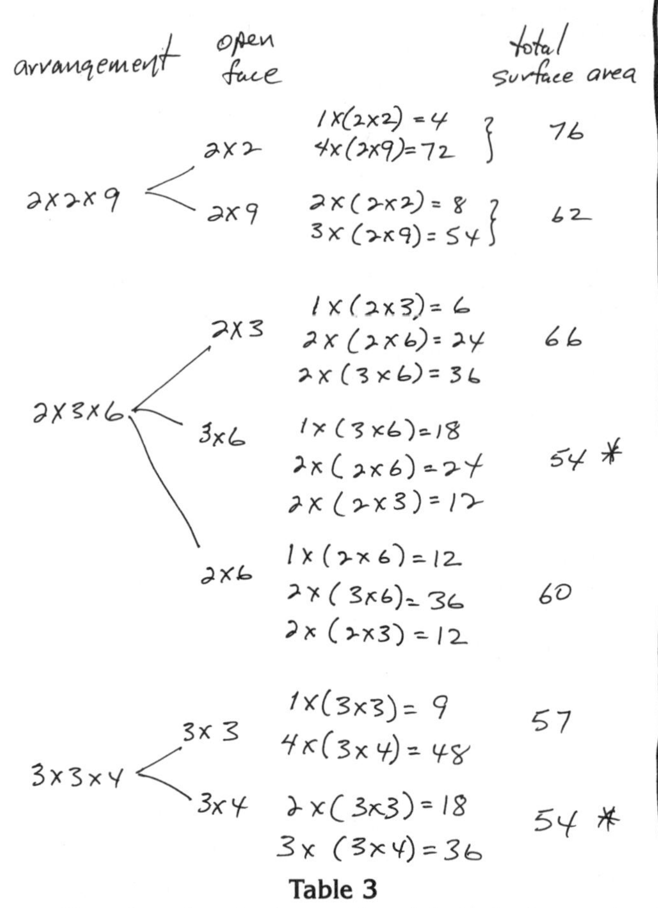

arrangement	open face		total surface area
2x2x9	2x2	1x(2x2) = 4 4x(2x9) = 72	76
	2x9	2x(2x2) = 8 3x(2x9) = 54	62
2x3x6	2x3	1x(2x3) = 6 2x(2x6) = 24 2x(3x6) = 36	66
	3x6	1x(3x6) = 18 2x(2x6) = 24 2x(2x3) = 12	54 *
	2x6	1x(2x6) = 12 2x(3x6) = 36 2x(2x3) = 12	60
3x3x4	3x3	1x(3x3) = 9 4x(3x4) = 48	57
	3x4	2x(3x3) = 18 3x(3x4) = 36	54 *

Table 3

Now that they had completed their design work, the two shippers went back to the puzzle. They wouldn't be able to make a final decision about which box to choose until they knew into which arrangement the puzzle pieces would fit. Since they had completed their design task and were caught up with their other work, they decided to spend some time trying to solve the puzzle themselves.

As the shippers tried putting the pieces together in various ways, they noticed that the six pieces were composed of three mirror image pairs. Hoping that this discovery might be a clue to solving the puzzle, they sorted the pieces into two sets so that each of them had one piece from each of the mirror image pairs. Within a few minutes of doing this, they both noticed a way to put their three pieces together to form half of a 2 x 2 x 9 arrangement (*Figure 4*). They had found one solution!

Figure 4

After putting the two identical halves together to form the 2 x 2 x 9 arrangement, they took them apart to think a bit more about what they had done. As they were examining the two pieces, they noticed a second way that they could be put together. This time the pieces formed a 2 x 3 x 6 arrangement *(Figure 5)*, giving them a second solution.

Figure 5

Now they wondered if there was yet another solution. Could it be possible to put the puzzle pieces together to form a 3 x 3 x 4 arrangement? They tried various other ways to separate the 2 x 3 x 4 arrangement into two identical pieces. As they looked at the puzzle from different sides, they suddenly noticed that it could be taken apart to form two identical pieces in yet another way. Moreover, they were excited to discover that the two pieces could be put back together in an arrangement different from the other two *(Figure 6)*. This time the two pieces formed the 3 x 3 x 4 arrangement they had been wondering about. This meant that the puzzle could be put together to form any of the three possible arrangements of 36 cubes. This also meant that the puzzle could be packed into any one of the boxes they had designed.

Figure 6

In the end, the shippers decided it would be fun to have three different boxes constructed for the puzzle: the 2 x 2 x 9 box with a 2 x 9 opening, the 3 x 3 x 4 box with a 3 x 4 opening, and the 2 x 3 x 6 box with a 3 x 6 opening. By offering the puzzle in three different boxes, a customer could order three of them, each with pieces arranged differently.

Background Article

Colored Squares
An Exploration

Perimeter and area of rectangles are important mathematical concepts generally introduced in the middle grades. An understanding of these concepts and their relationship to each other continue to be of importance throughout school mathematics.

Exploration

Consider a three by three square made of tiles *(Figure 1)*. Suppose the corner tiles are black, the tiles along each edge between the corners are gray, and the center tile is white. How many tiles are there of each color? Suppose we enlarge our square to a four by four, where once again the corner tiles are black, the tiles between the corners are gray, and the tiles in the center are white. Now how many tiles are there of each color? How many tiles would there be of each color if the square was enlarged to a five by five?

To answer the questions asked about these three squares, we may find ourselves simply counting the number of black, gray, and white tiles. This is especially true of the first two squares since they are pictured and it is very easy just to look at the figure and count the tiles. In the case of the five by five, however, it will probably be helpful to give at least some thought to what happened with the first two squares. As a last resort, it is easy to make a drawing like the ones for the first two squares. You should find that there are four black, 12 gray, and nine white tiles.

Figure 1

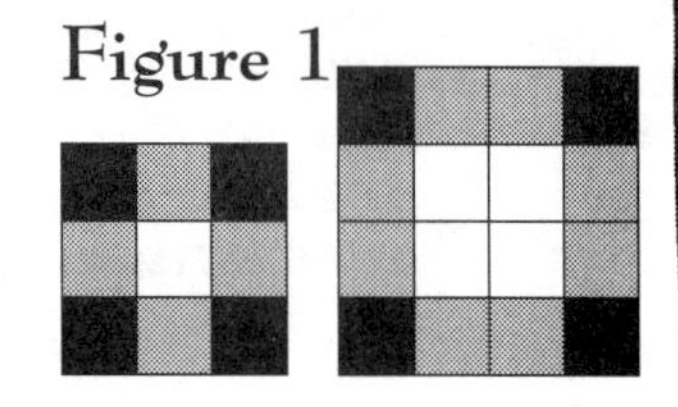

Suppose we jump ahead and ask about squares that are ten by ten or 50 by 50. How many tiles are there of each color? Do you see a pattern that can help you answer the question? Does the number of black tiles increase as the length of a side increases? How is the number of gray tiles related to the length of a side? What can you say about the relationship of the number of white tiles to the length of a side?

Figure 2

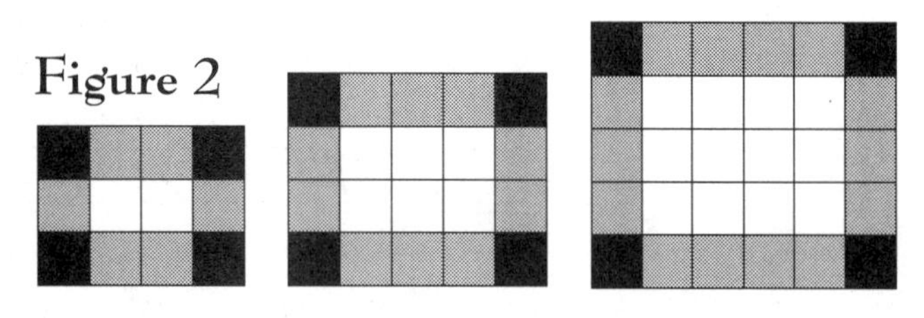

How would the counts of the various colored tiles change if they were arranged in the shape of a rectangle instead of a square? For example, suppose you had three by four, four by five, and five by six rectangles *(Figure 2)*. How many tiles would there be of each color? How many tiles of each color would there be if the rectangle were 15 by 16? How do the patterns for these rectangles compare to the patterns you found for squares?

Figure 3

Some Extensions

The first extension begins with a five by five square *(Figure 3)*. Note that the pattern of black, gray, and white tiles in the square has been changed. How many black, gray, and white tiles do you find for each square? How many of each would you find if the square was six by six or seven by seven *(Figure 4)*? How about if it was 20 by 20? What visual patterns do you see in the squares? Do you find number patterns that you can use to determine the number of tiles of each color for larger squares? How many would you find of each color if the square was 100 by100?

Figure 4

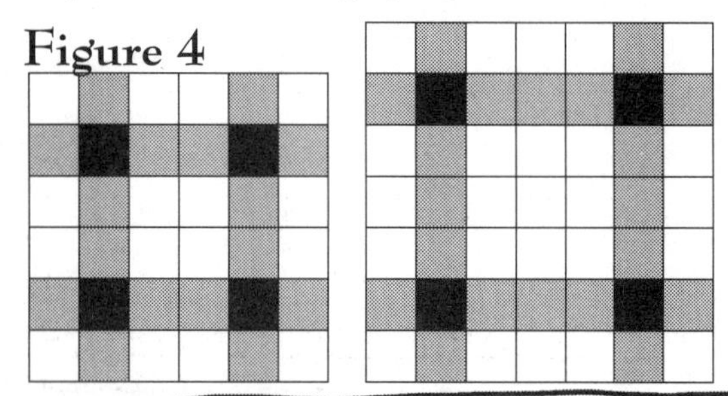

Figure 5

Figure 5 pictures a second arrangement of tiles that begins with a seven by seven square. *Figure 6* shows what the arrangement looks like for eight by eight and nine by nine squares. Do you see that while the number of black squares has changed in this arrangement, it still stays the same regardless of the size of the square? How many black, gray, and white tiles do you find for each square? Do you see more than one pattern in counting the number of gray and white tiles? How many would you find of each color if the square was 50 by 50?

Figure 6

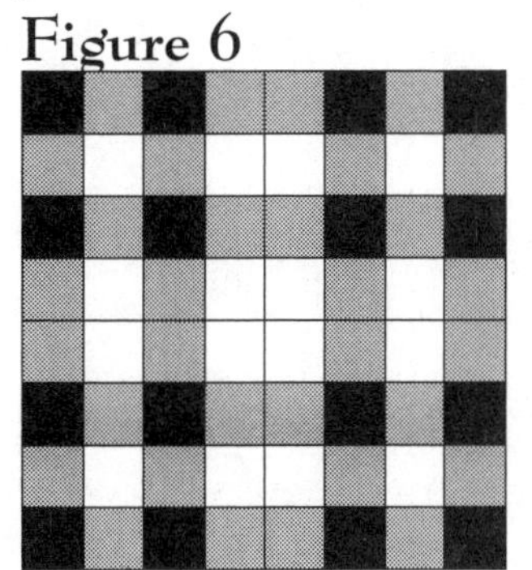

Discussion

What are some of the patterns that emerge from each of these four different arrangements of tiles? It's important to notice that it is the visual patterns that first strike us as we look at the different arrangements of the tiles. No one needs to ask us to notice the visual patterns in the arrangement of the tiles; it's as though we do it automatically. Even in the case of the two extensions where the arrangements are a little more complicated, finding patterns in the arrangements is almost irresistible. While our questioning was more about the number patterns in these arrangements, it is extremely important that we reflect on the role that the visual patterns play in determining the number patterns that emerge.

Table 1

Edge Length	Black Tiles	Gray Tiles	White Tiles	Total Tiles
3	4	4	1	9
4	4	8	4	16
5	4	12	9	25
6	4	16	16	36
7	4	20	25	49
8	4	24	36	64
9	4	28	49	81
10	4	32	64	100
20	4	72	324	400
50	4	192	2304	2500

To answer the questions posed, constructing some kind of table enables you to better see the emerging patterns. Some of your tables for the first square arrangement may have looked something like *Table 1*. Did you have a table where the entries for the gray tiles in the first arrangement looked like those in *Table 2?* Do these different ways of entering the numbers reflect the way in which the individual is thinking about the problem? Perhaps the entries in the *gray* column of *Table 1* are the result of simply counting the gray tiles, while the entries in that column of *Table 2* are the result of noticing that there are four edges and that the total number of gray tiles is four times the number of gray tiles in each edge. The number patterns in *Table 2* also show clearly that the number of gray on an edge is always two less than the length of the side of the edge.

Table 2

Edge Length	Black Tiles	Gray Tiles	White Tiles	Total Tiles
3	4	4x1	1x1	3x3=9
4	4	4x2	2x2	4x4=16
5	4	4x3	3x3	5x5=25
6	4	4x4	4x4	6x6=36
7	4	4x5	5x5	7x7=49
8	4	4x6	6x6	8x8=64
9	4	4x7	7x7	9x9=81
10	4	4x8	8x8	10x10=100
20	4	4x18	18x18	20x20=400
50	4	4x48	48x48	50x50=2500

Table 3 shows the number patterns that occur when the arrangement of tiles forms a rectangle, where the length is one unit greater than the width. The extra columns in *Table 3* show two different ways to think about the

Table 3

Edge Lengths	Black Tiles	Gray Tiles	White Tiles	Total Tiles
3x4	4	6=(2x1)+(2x2)	2=1x2	3x4=12
4x5	4	10=(2x2)+(2x3)	6=2x3	4x5=20
5x6	4	14=(2x3)+(2x4)	12=3x4	5x6=30
6x7	4	18=(2x4)+(2x5)	20=4x5	6x7=42
7x8	4	22=(2x5)+(2x6)	30=5x6	7x8=56
8x9	4	26=(2x6)+(2x7)	42=6x7	8x9=72
9x10	4	30=(2x7)+(2x8)	56=7x8	9x10=90
10x11	4	34=(2x8)+(2x9)	72=8x9	10x11=110
20x21	4	74=(2x18)+(2x19)	342=18x19	20x21=420
50x51	4	194=(2x48)+(2x49)	2352=48x49	50x51=2550

number of gray and white tiles. Again, the first column for each color reflects the result of simply counting the number of tiles of that color, while the second for gray shows a recognition that there are two pairs of equal edges. Similarly, the second column for white reflects thinking about the white tiles as a rectangular arrangement.

The visual and the number patterns both are more complex in the extensions. In the first extension, for example, while not the corner tile, there is a black tile corresponding to each corner. No matter the size of the square, there are only four black tiles. The corner tile of each of these squares is white and so there are at least four white tiles for each square. However, there are also white tiles along each edge and there are white tiles at the center of the square. Similarly, there are gray tiles between the black tiles and there are also two gray tiles on each edge next to the corner white tile. So how do we go about counting the tiles of different colors and how does our counting reflect our recognition of the various patterns? The multiple columns in *Table 4* reflect some of the possibilities. Again, notice the interrelationship of visual and number patterns.

Table 4

Edge Length	Black Tiles	Gray Tiles	White Tiles	Total Tiles
5	4	12=(4x2)+(4x1)	9=(4x1)+(4x1)+(1x1)	5x5=25
6	4	16=(4x2)+(4x2)	16=(4x1)+(4x2)+(2x2)	6x6=36
7	4	20=(4x2)+(4x3)	25=(4x1)+(4x3)+(3x3)	7x7=49
8	4	24=(4x2)+(4x4)	36=(4x1)+(4x4)+(4x4)	8x8=64
9	4	28=(4x2)+(4x5)	49=(4x1)+(4x5)+(5x5)	9x9=81
10	4	32=(4x2)+(4x6)	64=(4x1)+(4x6)+(6x6)	10x10=100
11	4	36=(4x2)+(4x7)	81=(4x1)+(4x7)+(7x7)	11x11=121
12	4	40=(4x2)+(4x8)	100=(4x1)+(4x8)+(8x8)	12x12=144
15	4	56=(4x2)+(4x12)	196=(4x1)+(4x12)+(12x12)	15x15=225
50	4	196=(4x2)+(4x47)	2401=(4x1)+(4x47)+(47x47)	50x50=2500

The second extension is even more complex than the first, and it's tempting simply to count the number of each color. However, there are some nice opportunities to connect number patterns to the more obvious visual patterns. Do you see how the number patterns in the gray and white columns of *Table 5* are related to the visual patterns in *Figures 5* and *6?*

This activity offers many opportunities to think about concepts and relationships associated with squares and rectangles. It requires thought about vertices, edges, perimeter, and area, and provides for an interplay between visual and number patterns related to these ideas. This would be an excellent set of activities to extend and supplement the teaching of area and perimeter.

Table 5

Edge Length	Black Tiles	Gray Tiles	White Tiles	Total Tiles
7	16	24=(4x4)+(4x2)x1	9=(4x1)+(4x1)+(1x1)	7x7=49
8	16	32=(4x4)+(4x2)x2	16=(4x1)+(4x2)+(2x2)	8x8=64
9	16	40=(4x4)+(4x2)x3	25=(4x1)+(4x3)+(3x3)	9x9=81
10	16	48=(4x4)+(4x2)x4	36=(4x1)+(4x4)+(4x4)	10x10=100
11	16	56=(4x4)+(4x2)x5	49=(4x1)+(4x5)+(5x5)	11x11=121
12	16	64=(4x4)+(4x2)x6	64=(4x1)+(4x6)+(6x6)	12x12=144
13	16	72=(4x4)+(4x2)x7	81=(4x1)+(4x7)+(7x7)	13x13=169
20	16	128=(4x4)+(4x2)x14	256=(4x1)+(4x14)+(14x14)	20x20=400
50	16	368=(4x4)+(4x2)x44	2116=(4x1)+(4x44)+(44x44)	50x50=2500

COLOR TILES

Topic
Patterns of geometric and algebraic growth

Key Question
How can you quickly determine the quantity of different colored tiles required to make a bordered tile pattern?

Learning Goals
Students will:
- extend patterns of different colored tiles as they grow sequentially,
- recognize numeric patterns, and
- represent these tiles graphically and symbolically.

Guiding Documents
Project 2061 Benchmarks
- *Logical connections can be found between different parts of mathematics.*
- *Mathematicians often represent things with abstract ideas, such as numbers or perfectly straight lines, and then work with those ideas alone. The "things" from which they abstract can be ideas themselves (for example, a proposition about "all equal-sided triangles" or "all odd numbers").*
- *Organize information in simple tables and graphs and identify relationships they reveal.*
- *Mathematical statements can be used to describe how one quantity changes when another changes. Rates of change can be computed from magnitudes and vice versa.*
- *Graphs can show a variety of possible relationships between two variables. As one variable increases uniformly, the other may do one of the following: always keep the same proportion to the first, increase or decrease steadily, increase or decrease faster and faster, get closer and closer to some limiting value, reach some intermediate maximum or minimum, alternately increase and decrease indefinitely, increase and decrease in steps, or do something different from any of these.*
- *The graphic display of numbers may help to show patterns such as trends, varying rates of change, gaps, or clusters. Such patterns sometimes can be used to make predictions about the phenomena being graphed.*

*NCTM Standards 2000**
- *Build new mathematical knowledge through problem solving*
- *Represent, analyze, and generalize a variety of patterns with tables, graphs, words, and, when possible, symbolic rules*
- *Relate and compare different forms of representation for a relationship*
- *Identify functions as linear or nonlinear and contrast their properties from tables, graphs, or equations*
- *Select and apply techniques and tools to accurately find length, area, volume, and angle measures to appropriate levels of precision*

Math
Measuring
  perimeter
  area
Algebra
  developing equations
  variables and constants
  graphing

Integrated Processes
Observing
Collecting and recording data
Recognizing patterns
Applying and generalizing

Materials
Student pages
Colored pencils, optional
Colored tiles or Hex-a-link cubes, optional

Background Information

Border tile patterns have an arrangement of tiles around the borders that remains constant. The length of the edges and the interior increase to accommodate different size rooms. As the patterns grow larger, the design in the corners remains constant, the edges grow in a linear pattern of constant additions, and the interior grows in area with the addition to its size growing greater each time.

By starting with a small pattern and enlarging it, one can count to find the patterns of growth. The corner patterns are constant because they are synonymous with the four points required to define a rectangle. The edges grow in the linear fashion because they are synonymous with the one-dimensional perimeter. The interior grows in a two-dimensional fashion as all area does.

The symbolic and representational displays of these patterns show the algebraic nature of their growth. The tiles confined to the corners remain constant and form a horizontal line of constant height on the graph. The tiles related to the edges grow by a constant factor and form a straight line with a constant slope. The tiles related to the interior have a squared relationship to the edge and form a parabolic curve of increasing steepness on the graph.

Exploring border tile patterns provides an opportunity to recognize and begin to understand the inter-relationships between visual, numeric, symbolic, and graphic patterns. It also provides a rich integration of the geometric concepts of measurement and dimensionality and their algebraic representations.

Management

1. Most students do well on this investigation at the representational level. Colored tiles or Hex-a-link cubes may be used as manipulatives if the students require a concrete experience.
2. The drawing, counting, and recording of tile patterns along with the initial numeric patterns can be completed in one period. Students should be given a longer period of time, such as overnight, to grapple with the large patterns and the algebraic solutions.

Procedure

1. Distribute *Color Tiles—1.* Focus student attention on the tile patterns by asking them to discuss what patterns they see in the sequence of growing tile arrangements. [The gray borders get longer, and the white interior square gets larger.]
2. Ask students to draw the next one or two arrangements in the sequence. Check to see that students are drawing them correctly.
3. Have students use the drawings to count and record the number of each color of tile in the arrangements.
4. Discuss with students the patterns they see in the chart. [There are always four blacks. There are four more grays each time. The white tiles grow by adding the next odd number in sequence, resulting in perfect square numbers.]
5. Have students use the patterns to complete the chart up to an arrangement with nine tiles on an edge. To verify the correctness, or if students are having trouble using a pattern, have them draw the arrangements.
6. Have students make a coordinate graph of the data using a different color for each color of tile.
7. Discuss with students what patterns they see in the graph. [The black tile line goes horizontally. The gray tile line is straight and goes up four tiles every time it moves to the right one tile length. The white tile line is a smooth curve that gets steeper and steeper.]
8. Ask students how their graphic patterns relate to the number patterns on the chart. [Black—always the same number, the line is always the same height; Gray—four more tiles each time, line goes up four tiles for each tile length it moves to the right; White—there are ever increasing numbers of tiles added each time, the line goes up steeper and steeper each time you go right one tile length.]
9. Ask students how the graphic patterns relate to what they see in the visual arrangements. [There is a constant number of blacks. Gray grows steadily. The white takes over the pattern very quickly.]
10. Allow time for students to determine how many tiles of each color are required for arrangements with 15 and 50 on an edge. (For 15, many students will draw or use an additive pattern to get the quantity. Both of these strategies become prohibitive with 50 on an edge. Students will need to grapple with the problem for some time.)
11. Have students discuss their solutions for arrangements with 15 and 50 on an edge. Include as discussion the strategies they used to get their solutions. (For the grays, students often recognize there are four edges, and the blacks always take up two of the tiles on an edge. They will then subtract those two from the edge length and multiply it by four to get the number of gray tiles. The border always makes the white square two narrower than the whole pattern. If two is subtracted from the edge length and the number is multiplied by itself, they get the number of white squares.)
12. Referring to the strategies and patterns utilized by the students, have the class come up with a method of determining the number of tiles of each color required for an arrangement of any length edge. Help the students translate their methods into algebraic equations. [black = 4, gray = 4(E-2), white = (E-2) x (E-2) = $(E-2)^2$, total = E x E = E^2]
13. Have students discuss how the equations are related to the graphic, numeric, and visual patterns.
14. Follow a similar procedure for the other two sequences of tile patterns.

Discussion

Discussion is included in *Procedure* for *Color Tiles—1.* Discussion for *Color Tiles—2,* and *Color Tiles—3* is similar. The solutions are displayed below.

Edge Length	Black Tiles	Gray Tiles	White Tiles	Total Tiles
3	4	4	1	9
4	4	8	4	16
5	4	12	9	25
6	4	16	16	36
7	4	20	25	49
8	4	24	36	64
9	4	28	49	81
10	4	32	64	100
20	4	72	324	400
50	4	192	2304	2500

Edge Length	Black Tiles	Gray Tiles	White Tiles	Total Tiles
3	4	4x1	1x1	3x3=9
4	4	4x2	2x2	4x4=16
5	4	4x3	3x3	5x5=25
6	4	4x4	4x4	6x6=36
7	4	4x5	5x5	7x7=49
8	4	4x6	6x6	8x8=64
9	4	4x7	7x7	9x9=81
10	4	4x8	8x8	10x10=100
20	4	4x18	18x18	20x20=400
50	4	4x48	48x48	50x50=2500

Edge Lengths	Black Tiles	Gray Tiles	White Tiles	Total Tiles
3x4	4	6=(2x1)+(2x2)	2=1x2	3x4=12
4x5	4	10=(2x2)+(2x3)	6=2x3	4x5=20
5x6	4	14=(2x3)+(2x4)	12=3x4	5x6=30
6x7	4	18=(2x4)+(2x5)	20=4x5	6x7=42
7x8	4	22=(2x5)+(2x6)	30=5x6	7x8=56
8x9	4	26=(2x6)+(2x7)	42=6x7	8x9=72
9x10	4	30=(2x7)+(2x8)	56=7x8	9x10=90
10x11	4	34=(2x8)+(2x9)	72=8x9	10x11=110
20x21	4	74=(2x18)+(2x19)	342=18x19	20x21=420
50x51	4	194=(2x48)+(2x49)	2352=48x49	50x51=2550

Extensions

1. Have students generate their own sequence of bordered tile patterns. They may want to use more than three colors. Have them determine the equations for the quantity of each color based on the pattern's edge length. Have them share their discoveries or trade with other students to solve.
2. Explore bordered tile patterns of rectangular shapes other than squares. Explore a sequence of rectangles that are similar in shape, or a sequence where the length and width each grow by one tile each time.

COLOR TILES-1

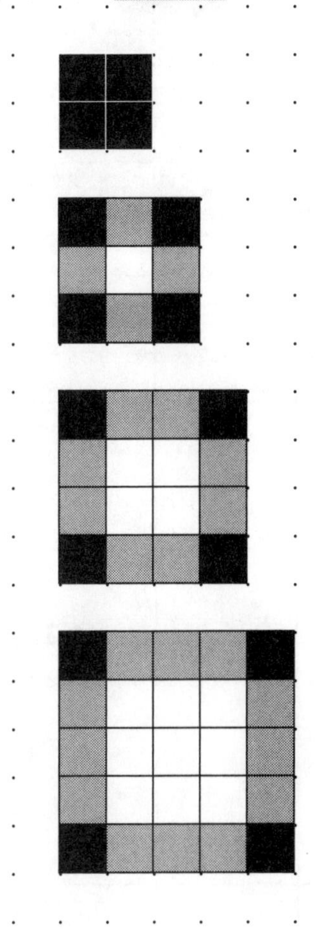

Edge Length	Black Tiles	Gray Tiles	White Tiles	Total Tiles
2				
3				
4				
5				
6				
7				
8				
9				
15				
50				
E				

COLOR TILES-1
NUMBER OF TILES
35
30
25
20
15
10
5
0
2
3
4
5
6
7
8
9
EDGE LENGTH

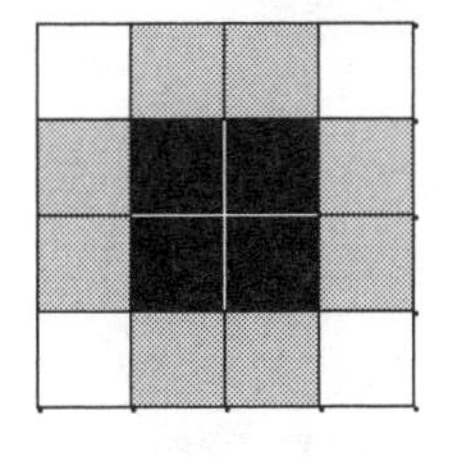

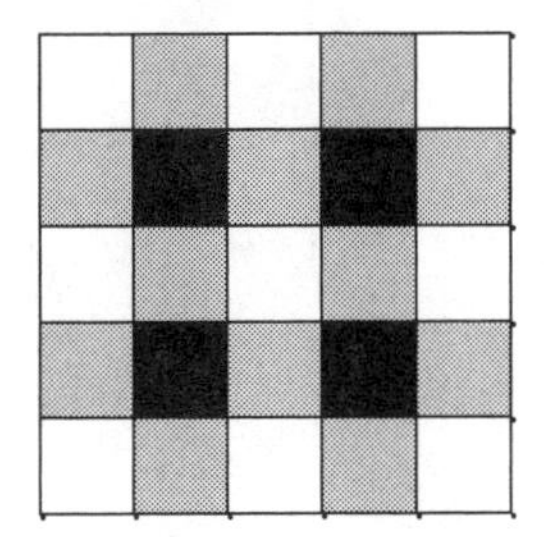

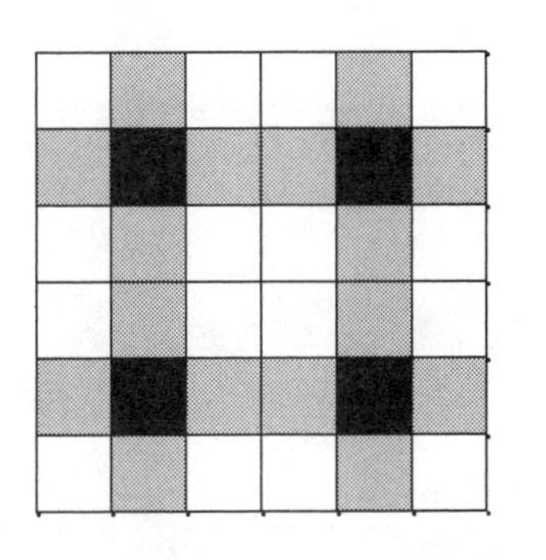

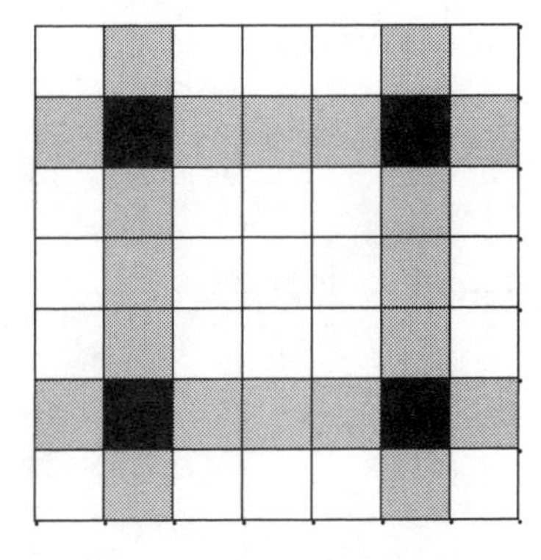

Edge Length	Black Tiles	Gray Tiles	White Tiles	Total Tiles
4				
5				
6				
7				
8				
9				
10				
11				
15				
50				
E				

COLOR TILES-2
70
60
50
40
30
20
10
0
NUMBER OF TILES
4
5
6
7
8
9
10
11
EDGE LENGTH

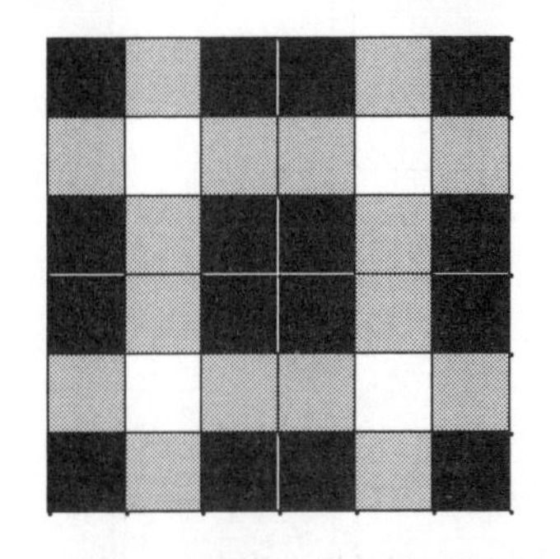

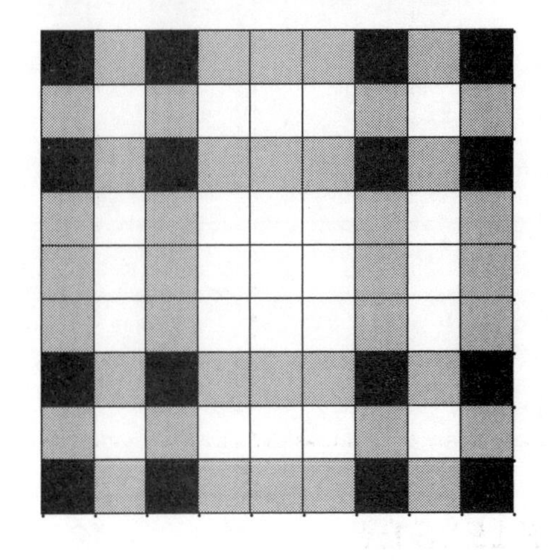

Edge Length	Black Tiles	Gray Tiles	White Tiles	Total Tiles
6				
7				
8				
9				
10				
11				
12				
13				
20				
50				
E				

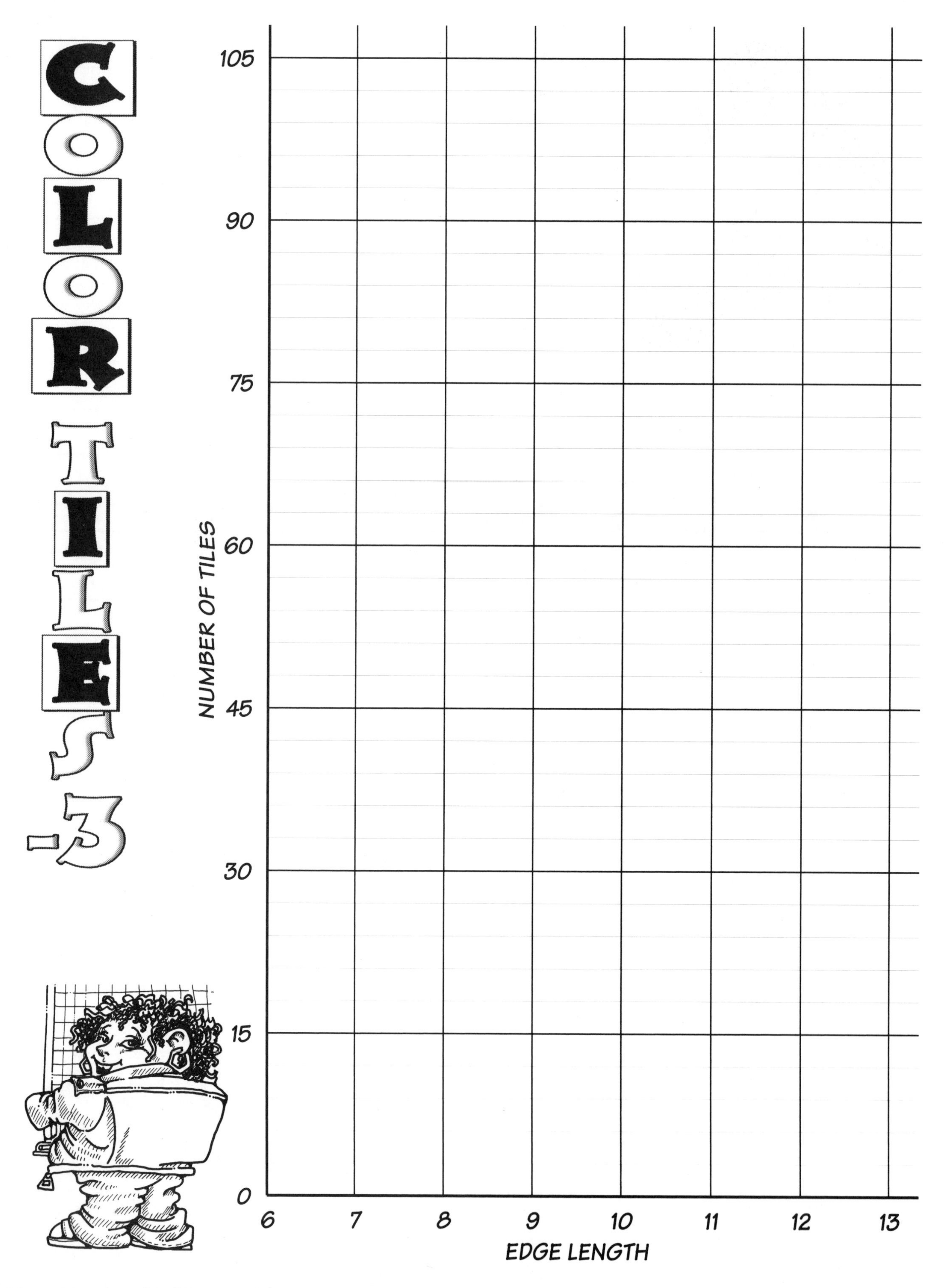
COLOR TILES -3
105
90
75
60
45
30
15
0
NUMBER OF TILES
6
7
8
9
10
11
12
13
EDGE LENGTH

Background Article

Painted Cubes

An Exploration

The richness of the cube as an object of exploration is both amazing and fascinating. While this is an article about the cube, it is also about spatial and number patterns, about dimension, and about length, area, and volume.

Exploration

Imagine that you are involved in the following activity. Twenty-seven wooden cubes are glued together to form a 3x3x3 cube. After the glue dries, the cube is given a coat of red paint. As you look at a face of the large cube, you see some of the faces of nine different small cubes. Examining the faces of these small cubes more closely, you notice that some of these cubes have faces that show only in the one face of the large cube, while some have faces that show in other faces of the large cube.

Suppose that you could take the large cube apart and examine each of the small cubes individually. What is the greatest number of painted faces that you would find for any one of the cubes? Would any of the cubes have no faces painted? Where in the large cube would you find cubes having different numbers of painted faces? How many of each of the differently painted cubes would be found in the 3x3x3 cube?

These questions are most easily answered if you have a set of 27 cubes available with which to construct a 3x3x3 cube. Or if cubes are not available, you can construct a 3x3x3 cube using one-inch or two-centimeter graph paper.

3x3x3 cube

For a 3x3x3 cube you will find that the greatest number of painted faces on any one of the small cubes is three. There are also cubes with exactly two painted faces, one painted face, and even one cube that has no painted faces. The cubes with three painted faces are at the eight corners of the large cube, those with two painted faces are along the edges between the corner cubes, those with one painted face at the center of each face of the large cube, and the one cube with no painted faces is out of sight in the interior of the large cube. *Table 1* shows the number of each kind of cube found in the 3x3x3 cube.

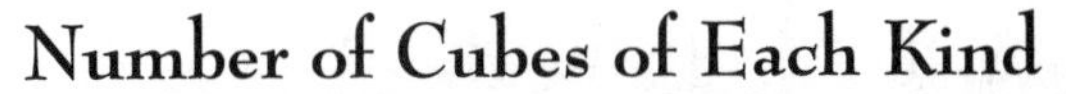

Number of Cubes of Each Kind

Table 1

Size of Large Cube	3 Faces Painted	2 Faces Painted	1 Face Painted	0 Faces Painted
3x3x3	8	12	6	1

How many small cubes of each kind would you have if you had started by painting a 4x4x4 cube instead of the 3x3x3 cube? How many with three-, two-, one-, and zero-painted faces? Are there more of each kind? Why or why not?

Did you notice that the number of cubes with three painted faces stayed the same? Every cube has exactly eight corners (vertices) and it is only these corner cubes that have three painted faces. Each edge now contains two cubes between the corner cubes. These are the ones with two painted faces and since there are 12 edges, there must be 24 of them. The center of each face is now a 2x2 square. These are the cubes with only one painted face and since there are four in each of the six faces, there must be 24 of them as well. Finally, how many cubes are in the interior of the 4x4x4 cube? These are the ones with no painted faces.

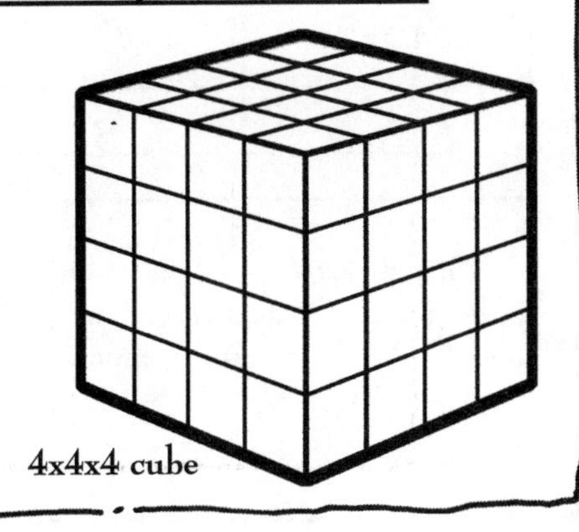

4x4x4 cube

Table 2 shows the number of small cubes of each kind for the 3x3x3 and the 4x4x4 cube. Can you complete the entries in the table for the 5x5x5 and 6x6x6 cubes? How about making the jump to the 10x10x10 or 15x15x15?

Number of Small Cubes of Each Kind

Table 2

Size of Large Cube	3 Faces Painted	2 Faces Painted	1 Face Painted	0 Faces Painted
3x3x3	8	12	6	1
4x4x4	8	24	24	8
5x5x5				
6x6x6				
10x10x10				
15x15x15				

Discussion

Some nice patterns begin to emerge in the first four lines of entries in *Table 2* that make it an easy jump to the 10x10x10. The eight corner cubes stay the same no matter the size of the large cube. For each of the 12 edges, the number of cubes with two painted faces is two less than the length of the edge. In each face of the large cube is a square arrangement of cubes with one painted face. This square has an edge that is two less than the edge of the larger cube. Finally, in the interior is a cube made up of unpainted small cubes. This cube, like the squares in each face, has an edge that is two less than the edge of the large cube. These patterns are even more evident if we rewrite them as has been done in *Table 3*.

Number of Small Cubes of Each Kind

Table 3

Size of Large Cube	3 Faces Painted	2 Faces Painted	1 Face Painted	0 Faces Painted
3x3x3	8	12x1	6x1	1
4x4x4	8	12x2	6x4	8
5x5x5	8	12x3	6x9	27
6x6x6	8	12x4	6x16	64

While the tables reveal the number patterns that emerge as we count the number of each kind of small cube, we have appealed to the visual pattern of these cubes. We have visualized corner cubes, edge cubes, a square of cubes within each face, and a cube of cubes in the interior. Observe the photo of the "exploded" 4x4x4 cube. Imagine a sequence of exploded views of successively larger cubes. Do you see the visual pattern that corresponds to the number pattern in the table?

An Alternate Exploration

There is another way to approach this problem that places greater emphasis upon the visual patterns. Make up several sets of 27 cubes that have been painted as shown in the 3x3x3 line of *Table 1.* Ask students to use the cubes to construct a 3x3x3 cube so that all faces of the cube appear painted. It is interesting to watch as individuals in the groups begin to examine the cubes. Some groups will begin immediately to build the cube, while others after examining the cubes will sort them according to the number of painted faces. As they sort and begin constructing the cube, their discussion begins to focus on the various kinds of small cubes and where these cubes will be located in the large cube.

While groups use different strategies, some taking longer than others, they are always successful in constructing the cube. After they have completed this task, immediately ask them to consider how many of each kind of small cube would be needed to construct a 4x4x4 cube. This turns out to be more difficult. Often it is not until the "how many of each kind of cube" question for the 4x4x4 cube is asked that they begin to think about how many of each kind were used for the 3x3x3. An even greater variety of strategies is generally seen as they attempt to answer this second question. Whatever the strategy, it generally involves an attempt to visualize the location and number of each kind of cube.

As students find answers to the "how many of each kind of cube" question for the 3x3x3 and 4x4x4 cubes, they often begin on their own to construct tables in which to record their findings. Not only do they make a record of the 3x3x3 and 4x4x4 cubes, but as if in anticipation of the next question, they begin to fill in the table for the 5x5x5 and so on.

Summary

The painted cube exploration is rich in opportunities to explore and generalize both visual patterns and corresponding number patterns. The location and the number of differently painted cubes call attention to the ideas of the vertices, edges, and faces of a cube. Unspoken in our exploration have been the concepts of dimensionality and length, and area and volume measurement. It was noted that the corner cubes with three faces painted are at the vertices which are zero dimensional. The cubes with two painted faces are along the edges which are one dimensional; the cubes with one painted face form a square which is two dimensional; the unpainted cubes form a cube which is three dimensional.

Topic
Dimensionality

Key Question
How can you determine the quantity of cubes painted on 0, 1, 2, and 3 sides in a large cube made of many smaller cubes?

Learning Goals
Students will:
- assemble a painted cube made of 27 smaller cubes,
- discover that the cubes have either 0, 1, 2, or 3 sides painted, and
- extend their experience to larger cubes and discover the patterns of growth in the quantity of cubes painted.

Guiding Documents
Project 2061 Benchmark
- *Calculate the circumferences and areas of rectangles, triangles, and circles, and the volumes of rectangular solids.*

*NCTM Standards 2000**
- *Build new mathematical knowledge through problem solving*
- *Represent, analyze, and generalize a variety of patterns with tables, graphs, words, and, when possible, symbolic rules*
- *Relate and compare different forms of representation for a relationship*
- *Identify functions as linear or nonlinear and contrast their properties from tables, graphs, or equations*
- *Select and apply techniques and tools to accurately find length, area, volume, and angle measures to appropriate levels of precision*
- *Recognize and use connections among mathematical ideas*

Math
Spatial visualization
Measuring
  dimensionality
    linear, area, volume
Algebra
  developing equations
  variables and constants

Integrated Processes
Observing
Collecting and recording data
Recognizing patterns
Applying
Generalizing

Materials
Wooden cubes (27 for each group puzzle, 216 for class models)
Paint, three colors
Paint brush
White glue
Graph paper
Isometric dot paper

Background Information
A classic math problem considers a painted cube made of many smaller cubes and asks how many smaller cubes there are of each configuration of painted sides. At this level, the problem is one of spatial visualization and counting. However, this situation provides a rich context for problems in patterning, dimensionality, growth, and algebra.

Students lacking skill in spatial visualization can be assisted by providing them with small cubic blocks. Simply building larger cubes from the blocks and counting them provides answers.

As answers from consecutively larger cubes are gained, numeric patterns emerge. There are always eight cubes that are painted on three sides; these comprise the corners of the larger cube. The quantity of cubes painted on two sides always goes up by 12. These cubes are found along the 12

edges. The cubes with only one side painted form the six faces of the cube. The quantity of cubes painted with one side is always a multiple of six. The other factor of the quantity is always the next perfect square number. The quantity of cubes with no painted sides is always a perfect cubic number.

Generalizing the numeric patterns highlights the algebraic concepts of constant and variable. As the patterns are translated into algebraic form, a connection develops between the abstraction of a linear equation and its linear equivalent, a line of cubes. Likewise the connection of X^2 is seen as a square on the face of a cube, and X^3 is seen as the interior cube.

Dimensions of Large Cube	Cubes with 3 Sides Painted	Cubes with 2 Sides Painted	Cubes with 1 Side Painted	Cubes with 0 Sides Painted	Total Number of Cubes
3x3x3	8	12	6	1	27
4x4x4	8	24	24	8	64
5x5x5	8	36	54	27	125
6x6x6	8	48	96	64	216
7x7x7	8	60	150	125	343
10 x 10 x 10	8	96	384	512	1000
50x50x50	8	576	13,824	110,592	125,000
LxLxL	8	12(L-2)	$6(L-2)^2$	$(L-2)^3$	L^3

As numerical and algebraic patterns are found in the cube, an understanding of dimensionality, its measure and growth, develops. The connection of dimension, measure, growth, and algebra are integrated in the cube. The eight points forming the corners have no dimension and are a constant. The lines of cubes forming the edges are different only in one dimension, length. The generalization of these perimeter pieces is represented with a linear expression. The squares of cubes forming the faces change in two dimensions. The surfaces of these areas are generalized with an expression to the power of two. The cube within grows in three dimensions. A cubic expression represents this growth in volume.

The richness of this elegant problem makes it worth much more attention than just a simple counting problem.

Management

1. This activity is best done over an extended period of time. Students should be given time to grapple with the problem outside of class. After completing the model on the first day *(Phase 1)*, students should be challenged to spend time finding solutions for larger puzzles *(Phase 2)*. Having students share the strategies and solutions for the larger puzzles should prepare them to spend some time grappling with the 10^3 and 50^3 puzzles *(Phase 3)*. During this time, the class models can be left out for student discussion. This extended approach provides for development of problem-solving skills, collaborative learning, and deeper understanding.
2. This activity provides for a wide range of understanding. A teacher should feel comfortable closing the activity when it has gone as far as appropriate for the students.
3. Although plain blocks may be used in larger cube puzzles, the students' literal understanding make painted blocks more beneficial. Before the activity, paint the cubes for the puzzle. A total of 27 cubes is required for each group. The painting list for each puzzles is:

Sides Painted	Number of Cubes
0	1
1	6
2	12
3	8
Total	**27**

4. Assembling the painted cube works well as a collaborative group puzzle. Allowing each student to handle only some of the cubes encourages better observations and discussion. The activity

can be done individually or at a station, but either alternative requires the *Procedure* to be modified.

5. Students find a physical model helpful for making the move to the abstract generalization. The teacher may choose to prepare specifically painted class models of cubes made from 27 (3^3), 64 (4^3), and 125 (5^3) cubes. For all three models, the cubes at the corners (three-sided) will be painted one color, the cubes at the edges (two-sided) will be painted a second color, and the cubes on only one face (one-sided) will be painted a third color. For ease in assembly, the two-sided cubes on each edge will be glued together in a line. The one-sided cubes on the face will be glued into flats. Those not painted will be glued into a solid cube. The cubes should be glued together before painting.

	1 like this	6 like this	12 like this	8 like this
3 x 3 x 3 Cube	1 cube used	6 cubes used	12 cubes used	8 cubes used
4 x 4 x 4 Cube	8 cubes used	24 cubes used	24 cubes used	8 cubes used
5 x 5 x 5 Cube	27 cubes used	54 cubes used	36 cubes used	8 cubes used

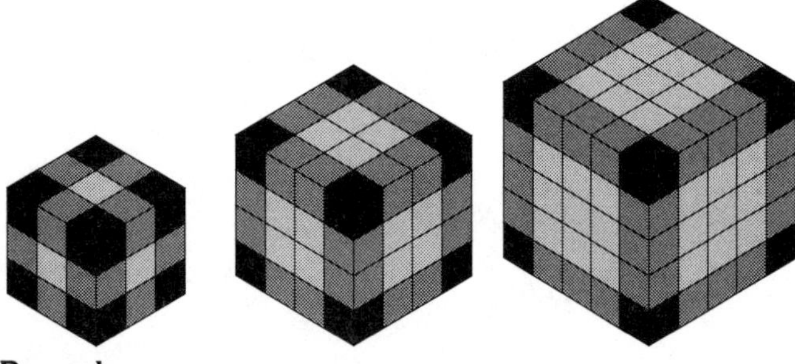

Procedure

Phase 1—Assembling the Painted Cube

1. Inform students that the objective of the puzzle is to assemble the blocks so a larger 3 x 3 x 3 cube is formed with all exposed sides painted. Inform them that each member in the group should get an equal amount of cubes and only that member should touch those pieces.
2. Have students work as a group to form the larger painted cube. Groups will follow different strategies to form their cubes but will go through the general pattern of sorting and counting the sides that are painted, determining where each type of painted cube would fit, and then following an organized system of assembly. Students will often have one or several surface squares unpainted and will need to reassemble the cube.
3. Distribute the chart and have students record the number of each type of cube in the chart.

Phase 2—Counting Larger Cubes

1. With constructed cubes in front of them, have students determine how many cubes of each type are required to make the next larger 4 x 4 x 4 cube. Make graph paper and isometric dot paper available to the students. Have students grapple with this problem. Many will begin to move the cubes they have to make space for new cubes and will count the spaces as cubes. Others will draw sketches. If no progress is being made, distribute the *Isometric Drawings* to help students.
2. When the groups have solutions, have them discuss how many of each type of painted cubes they counted and what strategies they used. The zero-sided and two-sided cubes create the most difficulty. The zero-sided cubes cause difficulty because they are completely hidden. The two-sided cause difficulty because many students work on a two-dimensional representation of one face. Often students will count the two-sided edge squares on one face and then multiply by six for the six faces, giving twice the correct amount. Many will follow a strategy similar to the following: three-sided, count corners; two-sided, count one edge and multiply by 12; one-sided, count face and multiply by six.
3. Using strategies they have developed, allow students time to determine the counts for the five-cubed, six-cubed, and seven-cubed puzzles.
4. When all groups have had ample time to get their counts, have the class verify that they have all gotten the correct counts. Then have them discuss what patterns they see in the chart. [Three-sided is always eight; two-sided go up by 12 each time; one-sided are multiples of six; zero-sided are cubed numbers; the totals are cubed numbers.]

Phase 3—Applying Patterns

1. Construct and display the painted three-cubed, four-cubed, and five-cubed class models. Have students discuss similarities and differences between the models. [The corners, edges, and faces are the same color on the models. There is the same number of corner pieces on all the models. There is one more edge cube each time. The square of face pieces gets one cube bigger

on an edge each time. The corner and edge colors are dominant on the first model and the face color is dominant on the last model.]

2. Disassemble each model and sort the types of pieces of each puzzle. (Observations are more easily recognized if the parts of each model are displayed in a long line of eight corners, 12 edges, six faces, and a cube. Then all three models can form a long line in a chalk tray or along a counter.) Have students discuss similarities and differences of parts of each of the models. [Every model has eight corners, 12 edges, six faces, and one unpainted cube. Each edge piece is one cube longer than the last model. Each face piece is one cube longer in two dimensions than the last model. Each cube is one cube longer in three dimensions than the last model.]
3. Allow students ample time to apply their observations to determining the number of each type of cube required to make a ten-cubed and 50-cubed puzzle. Have the students record their solutions on the chart. If they have difficulty, have students refer to the models and determine how long an edge piece was for a three-cubed, four-cubed, and five-cubed model. They should recognize that the edge is two less than the model size. Similar attention can be taken for the faces and unpainted cubes.
4. Have students discuss or demonstrate how they determined their solutions for each of the pieces. [For edges one might say, "I knew an edge piece was eight long because it's a corner shorter than the whole thing on both ends (10 - 2 = 8). Every model has 12 edges so I multiplied 12 times 8 and got 96 (12 x 8 = 96)."]
5. Have students generalize their patterns for determining the cubes for any puzzle of a given length (L).

Discussion

Phase 2

1. What strategy did you use to determine how many of a type of cube are in the puzzle? [make space, draw sketches, three-sided—count corners, two-sided—count one edge and multiply by 12, one-sided—count face and multiply by 6]
2. What patterns do you see in the columns of the chart? [Three-sided is always eight. Two-sided go up by 12 each time. One-sided are multiples of 6. Zero-sided are cubed numbers. The totals are cubed numbers.]

Phase 3

1. What similarities and differences do you notice about the parts of each of the models. [Every model has eight corners, 12 edges, six faces, and one unpainted cube. Each edge piece is one cube longer than the last model. Each face piece is one cube longer in two dimensions than the last model. Each cube is one cube longer in three dimensions than the last model.]
2. How did you determine the amount of cubes for each type of piece? [For edges one might say, "I knew an edge piece was eight long because it's a corner shorter than the whole thing on both ends (10 - 2 = 8). Every model has 12 edges so I multiplied 12 times 8 and got 96 (12 x 8 = 96)."]

Extension

Have students make exploded isometric views of models.

* Reprinted with permission from *Principles and Standards for School Mathematics,* 2000 by the National Council of Teachers of Mathematics.

Isometric Drawings

Isometric Drawings

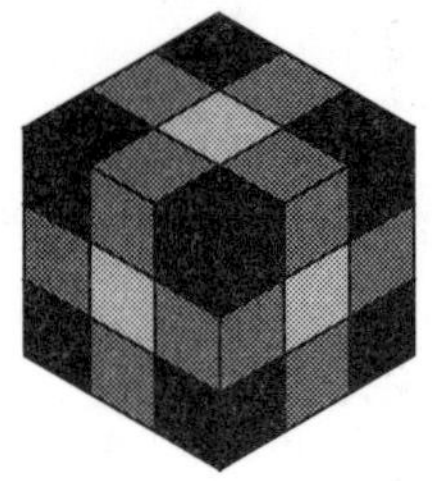

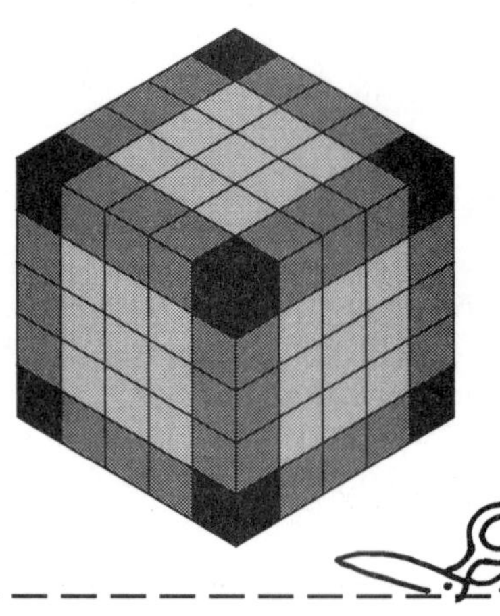

Isometric Drawings

Painted CUBES

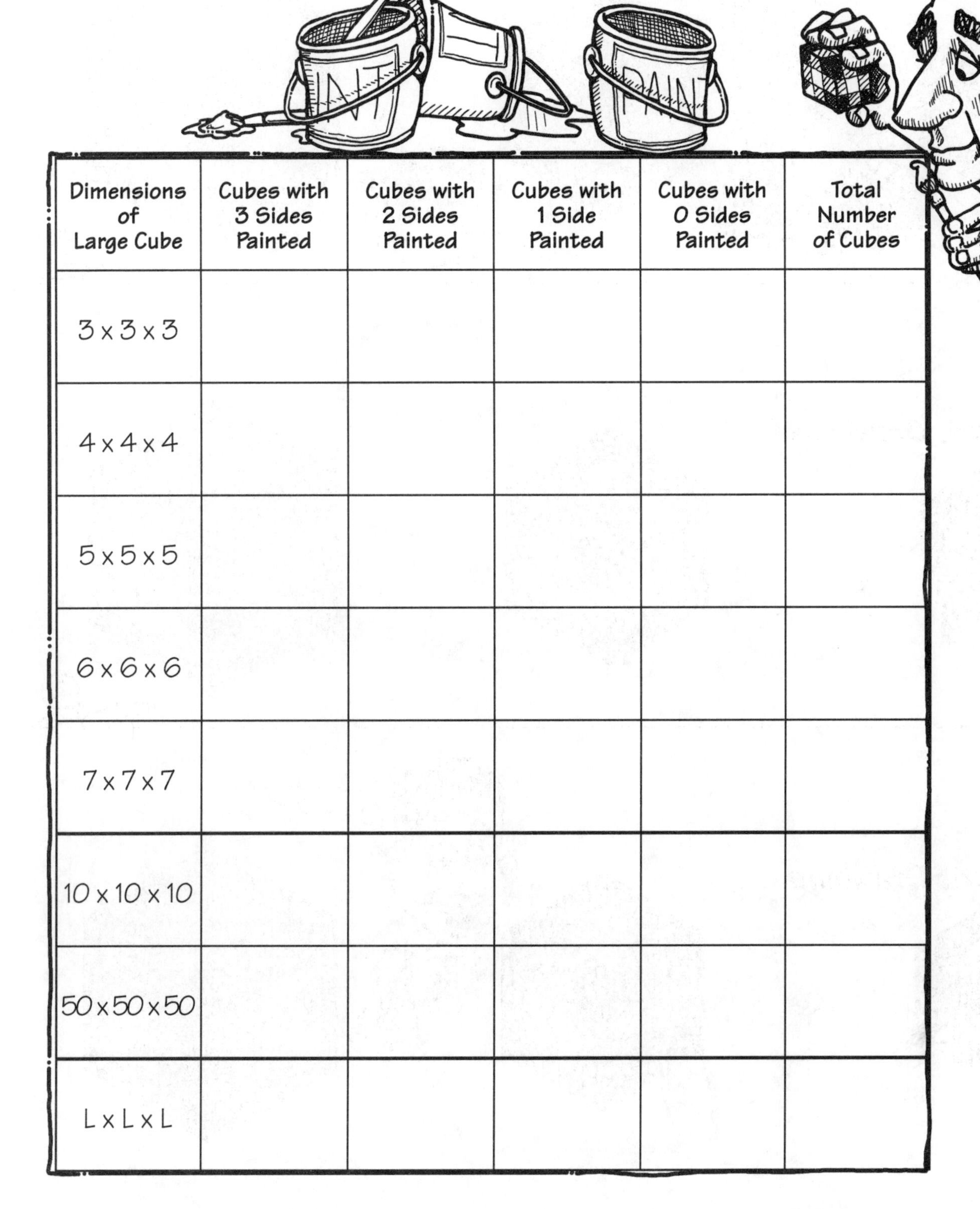

Dimensions of Large Cube	Cubes with 3 Sides Painted	Cubes with 2 Sides Painted	Cubes with 1 Side Painted	Cubes with 0 Sides Painted	Total Number of Cubes
3 x 3 x 3					
4 x 4 x 4					
5 x 5 x 5					
6 x 6 x 6					
7 x 7 x 7					
10 x 10 x 10					
50 x 50 x 50					
L x L x L					

Painted CUBES

Topic
Measurement: perimeter, surface area, volume

Key Question
How can you write a description of how to find a measurement of a cushioned tile once the dimension of the base component is determined?

Learning Goals
Students will:
- deepen their understanding of the meaning and purpose of perimeter, surface area, and volume;
- accurately calculate perimeter, surface area, and volume of rectangular regions and boxes; and
- generalize measurements of a figure algebraically when one dimension is unknown.

Guiding Documents
Project 2061 Benchmarks
- *Calculate the circumferences and areas of rectangles, triangles, and circles, and the volumes of rectangular solids.*
- *Organize information in simple tables and graphs and identify relationships they reveal.*

*NCTM Standards 2000**
- *Develop an initial conceptual understanding of different uses of variables*
- *Represent, analyze, and generalize a variety of patterns with tables, graphs, words, and, when possible, symbolic rules*
- *Select and apply techniques and tools to accurately find length, area, volume, and angle measures to appropriate levels of precision*
- *Recognize and apply mathematics in contexts outside of mathematics*

Math
Measurement
- perimeter
- surface area
- volume

Generalizing with expressions
Simplifying expressions

Integrated Processes
Observing
Collecting and recording data
Generalizing

Materials
AIMS Algebra Blocks

Background
This activity is based on a scenario of a company using a prototype to develop a new cushioned tile product. The product is made from sheets of dense foam one unit thick. The foam is cut into three different components that are assembled in different arrangements and are fused together to form a single tile. The base component is a large square of foam. The size of this base square is the debate within the company. As a prototype it would be best to consider the length as x, making the base component's dimensions $(1 \cdot x \cdot x)$. A second component is a trim strip. The trim strip's width is the same thickness as the foam, and it will be cut as long as the base component's edge. The strip's dimensions are $(1 \cdot 1 \cdot x)$. The third component is the tab. It is a cube with all edges as thick as the foam $(1 \cdot 1 \cdot 1)$.

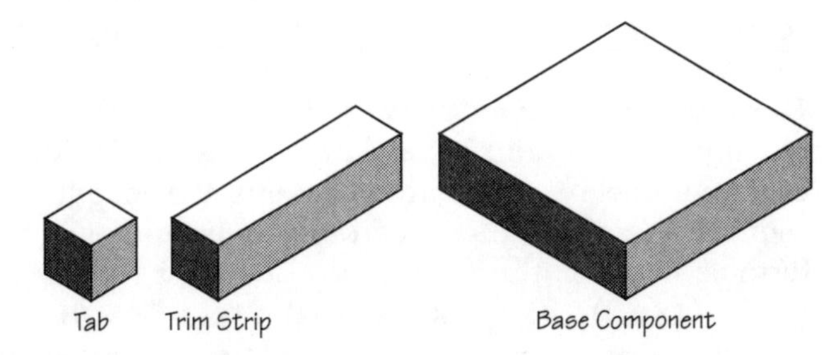

Many students have difficulty with perimeter, area, and volume because they have not internalized what type of measurement they are making. The scenario used in this activity has students visualize the measurement being made. They can imagine that fusing tape or binding tape will go around the perimeter of the tiles to hold them together. The cardboard that surrounds the tile packages is synonymous with area. Volume is seen as how much space the tiles take up in the package. By repeatedly calculating these measurements, students gain a familiarity with the measurements and processes for calculating the measurements. Being familiar with a method of calculation provides the understanding of generalizing these measurements in algebraic expressions.

The different methods used by students to calculate a measurement gives rise to different algebraic expressions. The study of these different expressions provides a rich experience for developing literacy in algebraic communication.

Consider the style example below:

To determine the perimeter, many students might start on a corner and go around adding pieces $(1 + x + 2 + x + 2 + x + 2 + x + 1)$. Students are quick to see that this can be written much more efficiently by combining the numbers and the variables $(4x + 8)$. Other students will approach the perimeter by recognizing there are four equal sides with the dimension $(x + 2)$. They will express the perimeter as $4(x + 2)$. This provides an opportunity to practice the distributive property to find the equivalence of $4(x + 2) = (4x + 8)$. When determining the area of the style, most students will count up each type of component so one base, four strips, and four tabs would be expressed $(x^2 + 4x + 4)$. This provides an opportunity to see whether the formula of length times width equals area works algebraically. Students can be asked to determine the length and width of the style $(x + 2)$ and then substitute it in the formula to see if it produces the equivalent expression $(x +2) \cdot (x +2) = (x^2 + 2x + 2x + 4) = (x^2 + 4x + 4)$

Looking at packaging the completed tiles provides even greater practice in relating numeric solutions to algebraic expressions and extending formulas. Consider the package example below:

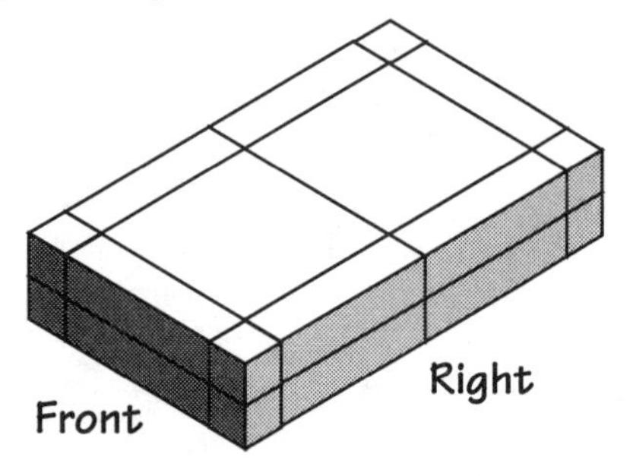

To find the surface area, students often count the pieces found on each face. The front would be $(2x + 4)$. The right side would be $(4x + 4)$. The top would be $(2x^2 + 6x + 4)$. Putting the areas of the three faces together and combining like terms would give an area of $(2x^2 + 12x + 12)$. Multiplying this area by two for the faces not seen in the drawing and applying the distributive property would provide the total surface area of $(4x^2 + 24x + 24)$. Students who are familiar with the formula $SA = 2\ (lw + lh + wh)$ could substitute the dimensions and solve the problem.

$SA = 2\ (lw + lh + wh)$
$SA = 2\ [(2x + 2) \cdot (x + 2) + (2x + 2) \cdot (2) + (x + 2) \cdot (2)]$
$SA = 2\ [(2x^2 + 6x + 4) + (4x + 4) + (2x + 4)]$
$SA = 2\ (2x^2 + 12x + 12)$
$SA = 4x^2 + 24x + 24$

Students generally attack the volume problem by counting the pieces and getting $4x^2 + 12x + 8$. Encouraging students to substitute the dimensions into the formula provides the same volume measurement and provides practice in meaningfully manipulating symbols.

$V = l \cdot w \cdot h$
$V = (2x + 2) \cdot (x + 2) \cdot (2)$
$V = (2x^2 + 6x + 4) \cdot (2)$
$V = (4x^2 + 12x + 8)$

For the concrete learner, determining the total edge length or length of binding tape required is completed by adding the lengths of all three bands of tape: the band around all the sides, the band around two sides and top and bottom, and the band around the front, back, top, and bottom.

$[(2x + 2) + (x + 2) + (2x + 2) + (x + 2)] + [(2 + (x + 2) + 2 + (x + 2)] + [(2 + (2x + 2) + 2 + (2x + 2)] = 12x + 24$

More abstract thinkers will recognize that it takes four segments of each dimension to get the total edge length and they will calculate it algebraically:

$4(2x + 2) + 4(x + 2) + 4(2) = 8x + 8 + 4x + 8 + 8 = 12x + 24$

Management

1. How students approach the situation of dealing with a prototype allows the teacher to assess what understanding of a variable a student has and an idea of the level of abstraction at which a student is working. Students who work out the measurement for base components of 3, 4, and 6 units and then struggle to write an algebraic expression are just beginning to understand the meaning of a variable and are strongly concrete thinkers. Students who want to work with the algebraic expression and then substitute in the values of 3, 4, and 6 for x are looking at the situation with abstract understanding and have developed an understanding of variable. The teacher will need to adjust the approach to best meet the needs of most of the students in the class.
2. The time required to complete this activity varies greatly with the experience students have had

with measurement and the level of abstraction at which they are working.

3. One set of Algebra Blocks for each group of four students is optimal. Working in small groups allows students to share understanding and clarify problems.

Procedure

1. Distribute the blocks to the students and discuss the scenario. Some time may need to be taken to make sure that students understand the concept of a prototype and that the dimension of the base unit (x) is not determined.
2. Direct students to determine the dimensions, surface area, volume, and total edge length of each type of algebra block when the base component is 3, 4, and 6 units long. [tab: 1 · 1 · 1, 6, 1, 12][trim strip: 1 · 1 · 3, 14, 3, 20; 1 · 1 · 4, 18, 4, 24; 1 · 1 · 6, 26, 6, 32 [base component: 1 · 3 · 3, 30, 9, 28; 1 · 4 · 4, 48, 16, 36; 1 · 6 · 6, 96, 36, 52]
3. Have students discuss and algebraically determine the dimensions, surface area, volume, and total edge length of each type of algebra block. [tab: 1 · 1 · 1, 1 sq. unit, 1 cu.unit, 6 units; trim strip: 1 · 1 · x, 4x + 2 square units, x cu.units, 4x + 8; base component: 1 · x · x, 2 x^2 + 4x sq.units, x^2 cu. units, 8x + 4 units]
4. Using the blocks, have the students construct each of the cushioned tile styles.
5. Referring to the sample tile, direct the students to determine what the perimeter and area would be for each style if the base unit's dimension were changed. For the algebraic expression, some time may need to be spent on combining like terms or using the distributive property to confirm that expressions are equivalent. (Refer to *Background Information.*)
6. Using the blocks, have the students construct each of the cushioned tile packages.
7. Referring to the sample packages, direct the students to determine what the surface area, volume, and total edge length would be for each package if the base unit's dimension was changed.

Discussion

1. How did finding the measurements for different base lengths help you generate the algebraic expressions in terms of x? (answers may vary, but many students find they count then add perimeters by base lengths and units and find area by counting squares, strips, and tabs)
2. How could you get the expression for the area of a style without counting the pieces? [record the length and width terms of x, and then multiply them using the distributive property]
3. Why is the algebraic expression using x units the most useful in dealing with a prototype? [Once the base dimension is determined, it can be substituted into the expression and the calculation only needs to be done once.]

Extensions

1. Have students construct their own tile styles and packages, record pictures of them and determine the algebraic expressions of the perimeter and area of the styles, and the surface area, volume, and total edge length of the packages. Direct the students to exchange their pictures and see if other students can determine the algebraic expressions for the measurements.
2. Have students record the length and width of different styles in terms of x, and then use them in the formulas $l \cdot w = A$ and $2(l + w) = P$ and see if they can generate the perimeter and area expressions for the each style.

You work for a company that is developing a new cushioned floor tile that will be marketed around the country. It is still in its prototype stage and management has come to you for your expertise in measurement.

The tile is made from a dense thick form. There are three different components that are assembled in different arrangements and are fused together to form a single tile. The base component is a large square of foam. A second component is a trim strip. The trim strip's width is the same thickness as the foam, and it will be cut as long as the base component's edge. The third component is the tab. It is a cube with all edges as thick as the foam.

The company has yet to determine the size of the basic component. The engineering department has made some proposals, but the marketing department will decide what will sell best. They have given you the prototype material to use to help you in your calculations.

The company will use the following components to make cushioned tiles. Use your prototype materials and determine the measurement of each piece for each of the dimensions given for the base component's length.

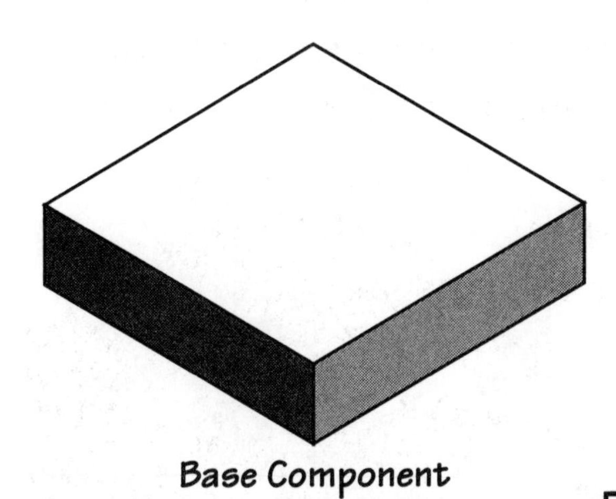

Base Component

Trim Strip

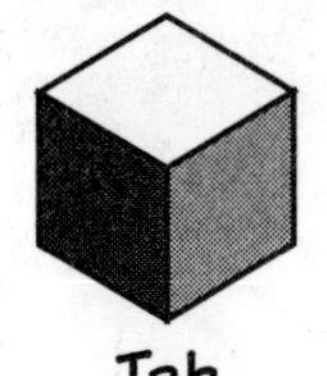

Tab

		PROPOSED DIMENSION OF BASE COMPONENT			
		3 units	4 units	6 units	X units
Surface Area	Tab				
	Strip				
	Base				
Volume	Tab				
	Strip				
	Base				
Total Edge Length	Tab				
	Strip				
	Base				

CUSHIONED
TILE
Styles
The company is considering the following styles of cushioned tiles. Construct them from your prototype materials and determine their dimensions as requested.
A
B
C
D
E
F
G
H
RESEA
DEPA
Search

Perimeter

Binding tape surrounds the arrangement of components to create a single tile. The company has not yet determined what size to make the base component's length. Calculate the tape needed for each style with each proposed base length.

		PROPOSED DIMENSION OF BASE COMPONENT			
		3 units	4 units	6 units	X units
STYLES	A				
	B				
	C				
	D				
	E				
	F				
	G				
	H				

Area

Interior designers need to know how much area each tile will cover. The company has not yet determined what size to make the base component's length. Calculate the area covered by each style for each proposed base length.

		PROPOSED DIMENSION OF BASE COMPONENT			
		3 units	4 units	6 units	X units
STYLES	A				
	B				
	C				
	D				
	E				
	F				
	G				
	H				

CUSHIONED
TILE
Packages
The company is considering the following packages of cushioned tiles. Construct them from your prototype materials and determine their dimensions as requested.
SMILE
Manager
#1
A
B
C
D
E

Cushioned Tile Packages

Surface Area

Protective cardboard covers each side of the package. Determine the number of square units of cardboard the fabricators have to order to cover each package.

PACKAGES	PROPOSED DIMENSION OF BASE COMPONENT			
	3 units	4 units	6 units	X units
A				
B				
C				
D				
E				

Volume

Shippers need to determine how many packages can be put in each truck. Determine the cubic units of volume each package fills.

PACKAGES	PROPOSED DIMENSION OF BASE COMPONENT			
	3 units	4 units	6 units	X units
A				
B				
C				
D				
E				

Total Edge Length

Binding tape holds the package and cardboard together. It is wrapped around each package in all three directions. Calculate how much tape each package requires.

PACKAGES	PROPOSED DIMENSION OF BASE COMPONENT			
	3 units	4 units	6 units	X units
A				
B				
C				
D				
E				

Topic
Measurement and Algebraic Thinking

Key Question
How can you write a description of how to find a measurement of a box set once the dimension of the base component is determined?

Learning Goals
Students will:
- deepen their understanding of the meaning and purpose of perimeter, surface area, and volume;
- accurately calculate length, surface area, and volume of rectangular regions and boxes; and
- generalize measurements of a figure algebraically when one dimension is unknown.

Guiding Documents
Project 2061 Benchmarks
- *Calculate the circumferences and areas of rectangles, triangles, and circles, and the volumes of rectangular solids.*
- *Organize information in simple tables and graphs and identify relationships they reveal.*

*NCTM Standards 2000**
- *Develop strategies to determine the surface area and volume of selected prisms, pyramids, and cylinders*
- *Represent, analyze, and generalize a variety of patterns with tables, graphs, words, and, when possible, symbolic rules*
- *Develop an initial conceptual understanding of different uses of variables*

Math
Measurement
  length
  surface area
  volume
Generalizing with expressions
Simplifying expressions

Integrated Processes
Observing
Collecting and recording data
Generalizing

Materials
AIMS Algebra Blocks

Background Information
This activity is based on a scenario of a company using a prototype to develop a set of blocks made of cushion material. The product is made from sheets of dense foam one unit thick. The foam is cut into four different components that are assembled to form a rectangular box that will be sold as a set. The size of the components will be based on the large cube of foam. The size of this cube is the debate within the company. As a prototype, it would be best to consider the length as x, making the cube's dimensions $(x \cdot x \cdot x = x^3)$. A second component is a flat square piece. The flat is one unit thick and will be the same length and width as the cube. The flat's dimensions would be $(1 \cdot x \cdot x = x^2)$. The third component is the column. The column's width is the same thickness as the foam, and it will be cut as long as the cube's edge. The column's dimensions are $(1 \cdot 1 \cdot x = x)$. The fourth component is the unit. It is a cube with all edges as thick as the foam $(1 \cdot 1 \cdot 1 = 1)$.

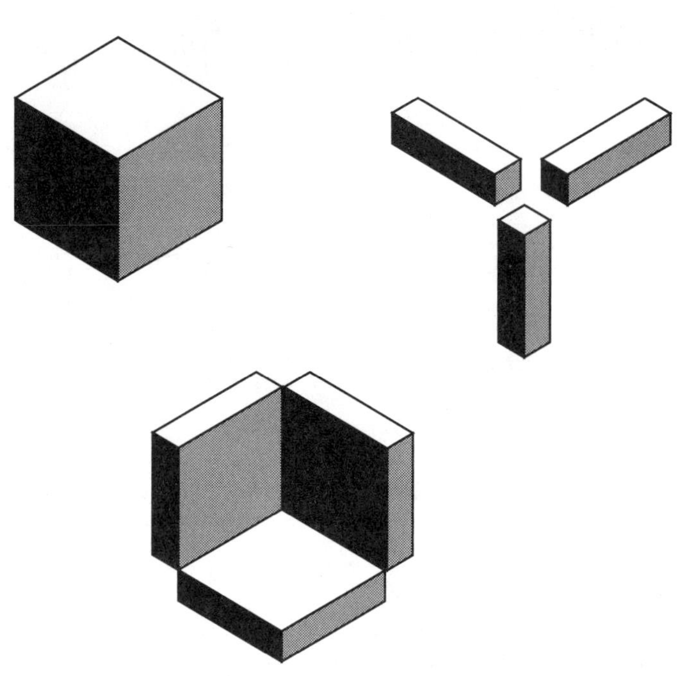

Many students have difficulty with surface area and volume because they have not internalized what type of measurement they are making. The scenario used in this activity has students visualize the measurement being made. The cardboard that surrounds the block set is synonymous with area. Volume is seen as how much space the blocks take up in the package. By repeatedly calculating these measurements, students gain a familiarity with the measurements and processes for calculating them. Being familiar with a method of calculation provides the understanding of generalizing these measurements in algebraic expressions.

The different methods used by students to calculate a measurement gives rise to different algebraic expressions. The study of these different expressions provides a rich experience for developing literacy in algebraic communication. Consider the package example below:

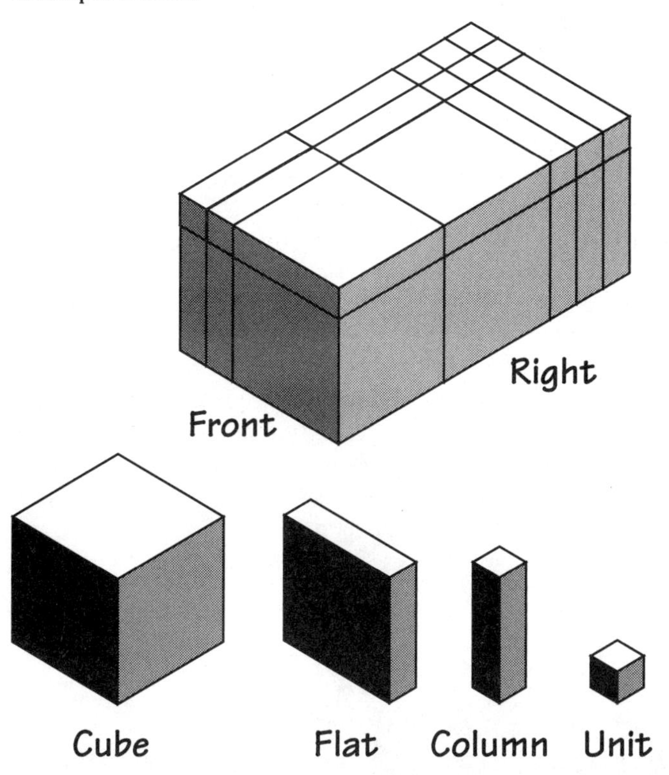

To find the surface area, students often count the pieces found on each face. The front would be ($x^2 + 3x + 2$). The right side would be ($2x^2 + 5x + 3$). The top would be ($2x^2 + 7x + 6$). Putting the areas of the three faces together and combining like terms would give an area of ($5x^2 + 15x + 11$). Multiplying this area by two for the faces not seen in the drawing and applying the distributive property would provide the total surface area of ($10x^2 + 30x + 22$). Students who are familiar with the formula SA = 2 (lw + lh + wh) could substitute the dimensions and solve the problem.

$SA = 2\ (lw + lh + wh)$
$SA = 2\ [(x + 2) \cdot (x + 1) + (2x + 3) \cdot (x+1) + (x + 2) \cdot (2x+3)]$
$SA = 2\ [(x^2 + 3x + 2) + (2x^2 + 5x + 3) + (2x^2 + 7x + 6)]$
$SA = 2\ (5x^2 + 15x + 11)$
$SA = 10x^2 + 30x + 22$

Students generally attack the volume problem by counting the pieces and getting $2x^3 + 9x^2 + 13x + 6$. Encouraging students to substitute the dimensions into the formula provides the same volume measurement and provides practice in meaningfully manipulating symbols.

$V = l \cdot w \cdot h$
$V = (2x + 3) \cdot (x + 2) \cdot (x + 1)$
$V = (2x^2 + 7x + 6) \cdot (x + 1)$
$V = (2x^3 + 9x^2 + 13x + 6)$

Management

1. How students approach the situation of dealing with a prototype allows the teacher to assess a student's understanding of a variable and an idea of the level of abstraction at which a student is working. Students who work out the measurements with numbers and then struggle to write an algebraic expression are just beginning to understand the meaning of a variable and are strongly concrete thinkers. Students who want to work with the algebraic expression and then substitute in the values of 3, 4, and 6 for x are looking at the situation with abstract understanding and have developed an understanding of variable. The teacher will need to adjust the approach to best meet the needs of most of the students in the class.
2. The time required to complete this activity varies greatly with the experience students have had with measurement and the level of abstraction at which they are working.
3. One set of Algebra Blocks for each group of four students is optimal. Working in small groups allows students to share understanding and clarify problems.

Procedure

1. Distribute the blocks to the students and discuss the scenario. Some time may need to be taken to make sure that students understand the concept of a prototype and that the dimension of the base unit (x) is not determined.
2. Direct students to determine the dimensions, surface area, and volume of each type of Algebra Block when the cube's length is 3, 4, and 6 units long. [unit: $1 \cdot 1 \cdot 1$, 6] [column: $1 \cdot 1 \cdot 3$, 14, 3; $1 \cdot 1 \cdot 4$, 18, 4; $1 \cdot 1 \cdot 6$, 26, 6] [flat: $1 \cdot 3 \cdot 3$, 30, 9; $1 \cdot 4 \cdot 4$, 48, 16; $1 \cdot 6 \cdot 6$, 96, 36] [cube: $3 \cdot 3 \cdot 3$, 54, 27; $4 \cdot 4 \cdot 4$, 96, 64; $6 \cdot 6 \cdot 6$, 216, 216]
3. Have students discuss and algebraically determine the dimensions, surface area, and volume of each type of Algebra Block. [unit: $1 \cdot 1 \cdot 1$, 1 sq. unit, 1 cu. unit; column: $1 \cdot 1 \cdot x$, $4x + 2$ sq. units, x cu. units; flat: $1 \cdot x \cdot x$, $2x^2 + 4x$ sq. units, x^2 cu. units; cube: $x \cdot x \cdot x$, $6x^2$ sq. units, x^3 cu. units]
4. Have each group make a rectangular box out of the number of components listed for the basic unit on the record page.
5. Referring to the group's box, invite each student to draw the shape of a net for that box and include the outline of each block on the face of the net.
6. Have the students determine and record what the surface area and volume would be for the box if the cube's dimensions were changed. For the algebraic expression, some time may need to be spent on combining like terms or using the distributive property to confirm that expressions are equivalent. Refer to *Background Information.*
7. If more practice is needed, have students follow the same procedure for the deluxe set of blocks.

Discussion

1. How did finding the measurements for different lengths for the cube help you generate the algebraic expressions in terms of x? (answers may vary, but many students find area by counting square faces, column faces, and unit faces)
2. How could you get the expression for the area of a face of a box without counting the pieces? [record the length and width terms of x, and then multiply them using the distributive property]
3. Why is the algebraic expression using x units the most useful in dealing with a prototype? [Once the base dimension is determined, it can be substituted into the expression and the calculation only needs to be done once.]

Extension

Have students construct structures out of the blocks in one of the sets. Tell them to make an isometric drawing of their tower. Then ask them to determine the maximum height and width of their tower when the cubes dimensions are 3, 4, 6, or x units long.

* Reprinted with permission from *Principles and Standards for School Mathematics,* 2000 by the National Council of Teachers of Mathematics. All rights reserved.

SOFT BLOCS

You work for ACME Toy Company. They are developing a new set of blocks made of foam material. Young children will be able to build a tall tower that will not hurt them when it tumbles.

The toy is still in the prototype testing stage and the company is undecided as to what size to make the blocks. The blocks need to be ready for the holiday rush so the company is coming to you as a measurement expert to get advice on packaging in order to be ready when the size decision is made.

The blocks are made of soft foam that comes in large sheets one unit thick. There will be four types of blocks made from the foam. The largest block is a large cube. The rest of the blocks will be based on the size of this large cube's length. The second block is a flat square piece. The square flat is one unit thick like the foam, but the square face will be the same size as the cube. The column is the third block. It will be one unit wide by one unit thick and will be the length of the cube. The last block is a unit cube that has the same dimensions as the unit thickness of the foam sheet from which it is cut.

Your job will be to determine how to efficiently package the sets, design a net for the each set's box, determine the amount of cardboard needed for the box, and calculate the space taken up in the delivery truck. If you get all this work done, the company would like you to design some towers that can be made from the set.

SOFT BLOCS

The company will use the following blocks to make up the sets of Soft Blocs. Use your prototype materials and determine the measurements of each piece for each dimension given to the cube.

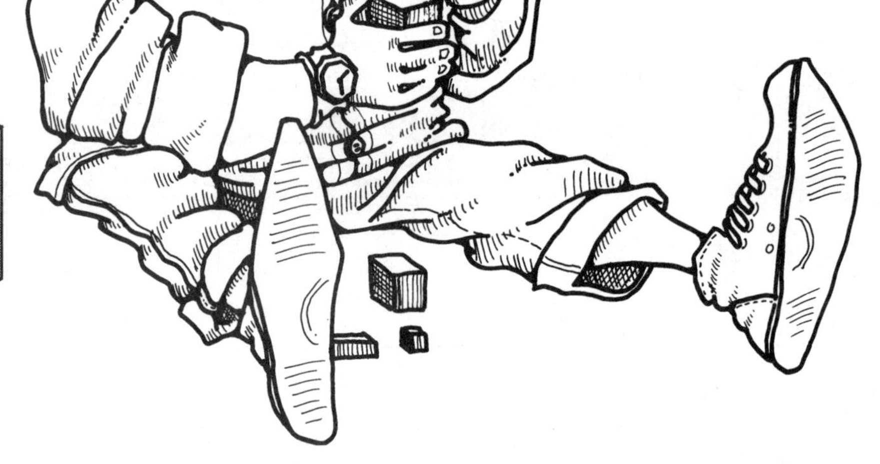

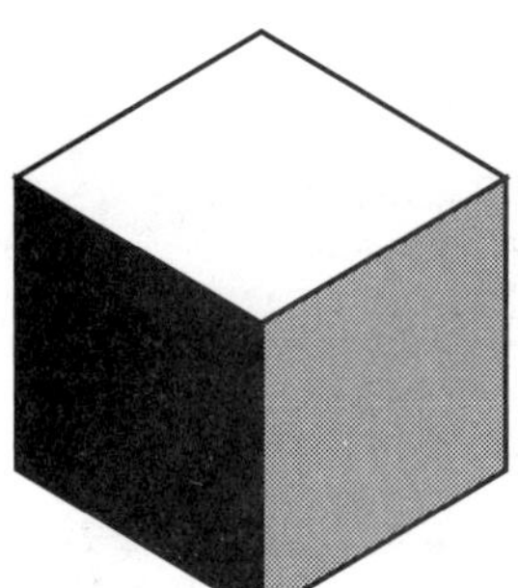

Cube

Flat

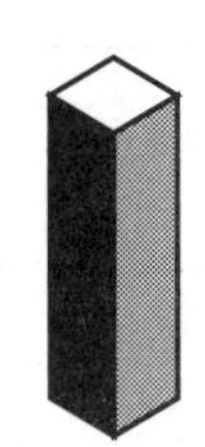

Column

Unit

		PROPOSED DIMENSION OF CUBE			
		3 units	4 units	6 units	X units
SURFACE AREA	Unit				
	Column				
	Flat				
	Cube				

VOLUME	Unit				
	Column				
	Flat				
	Cube				

1. Make a rectangular solid with all the components of a block set.
2. Draw the shape of the net that would surround the block set. Include the outline of each block on the face.
3. Record the dimensions and volume of the box for each possible size.
4. Determine the surface area of each face for each possible size.

The Basic Soft Bloc Set:
2 - Cubes
9 – Flats
13 – Columns
6 – Units

The Deluxe Soft Bloc Set:
4 - Cubes
16 – Flats
20 – Columns
8 – Units

		PROPOSED DIMENSION OF CUBE			
		3 units	4 units	6 units	X units
Dimensions					
Volume					
SURFACE AREA OF EACH FACE	Top Face				
	Bottom Face				
	Front Face				
	Back Face				
	Right Face				
	Left Face				
Total Surface Area					

SOFT BLOCS

ACME Toy Co. Inc.

Box Net Drawn by:

KEY
Outlines of blocks

Make a structure of your own design from a Soft Bloc set. Make a isometric sketch of the structure and determine the maximum width and height of the structure.

	PROPOSED DIMENSION OF CUBE			
	3 units	4 units	6 units	X units
Dimensions				
Volume				

Topic
Measurment and Algebraic Thinking

Key Question
How can you work backwards from the cost of a window and one of its dimensions to determine the other dimensions of the window?

Learning Goals
Students will:
1. practice using geometric formulas in translating situations, and
2. solve equations with one variable.

Guiding Documents
Project 2061 Benchmark
- *When mathematicians use logical rules to work with representations of things, the results may or may not be valid for the things themselves. Using mathematics to solve a problem requires choosing what mathematics to use; probably making some simplifying assumptions, estimates, or approximations; doing computations; and then checking to see whether the answer makes sense. If an answer does not seem to make enough sense for its intended purpose, then any of these steps might have been inappropriate.*

*NCTM Standards 2000**
- *Use symbolic algebra to represent situations and to solve problems, especially those that involve linear relationships*
- *Develop and use formulas to determine the circumference of circles and the area of triangles, parallelograms, trapezoids, and circles and develop strategies to find the area of more complex shapes*

Math
Measurement
 area
Algebraic thinking
 writing equations
 solving equations

Integrated Processes
Observing
Applying
Generalizing

Materials
AIMS Algebra Blocks, optional

Background Information
This practice activity is provided for students to apply and reinforce what they know about area and working backwards. Five models of windows are shown where the width (w) dimensions vary. With a constant of $2.00/ft², the students can set up an equation to the unit cost and work backwards to solve for the width. Below is the chart with an equation and the solution.

Window Type	Unit $	Equation	Solution
Transom	9.00	2(1.5w)=9	w=3
Transom	15.00	2(1.5w)=15	w=5
Diamond	18.00	2(w2)=18	w=3
Diamond	12.50	2(w2)=12.5	w=2.5
Louvered(5-panel)	30.00	2(5w)=30	w=3
Louvered(4-panel)	20.00	2(4w)=20	w=2.5
Double Hung	36.00	2(2w2)=36	w=3
Double Hung	25.00	2(2w2)=25	w=2.5
Picture	64.00	2(w2+4w)=64	w=4
Picture	42.00	2(w2+4w)=42	w=3

Some students will guess at the width and then calculate the cost of the window to see if their guess was correct. Many students will work backwards intuitively and will need to have help translating into algebraic terms what they thought. They will talk about finding the cost (c) by determining the area (l·w) and then doubling it because it cost two dollars a square foot. Algebraically this is written:

$$2(l \cdot w) = c$$

Substituting what is known in the first problem the equation is:

$$2(1.5 \cdot w) = 9$$

Students will talk about dividing the nine dollar cost to determine the area.

$$\frac{2(1.5 \cdot w)}{2} = \frac{9}{2}$$

$$1.5 \cdot w = 4.5$$

By dividing the area by the length you get the width.

$$\frac{1.5 \cdot w}{1.5} = \frac{4.5}{1.5}$$

$$w = 3$$

Management
1. *Cushion Tile Develepment* or *Acme Toy Co., Inc.* should precede this activity.
2. The AIMS Algebra Blocks can be used as manipulatives to help students work with the problems.

Procedure
1. Distribute the student page and make sure students understand the context.

2. Allow students to begin working on the problem. Provide Algebra Blocks for students to use. After students have had some time to grapple with the problem individually, allow them to discuss their insights, strategies, and problems with a partner or group.
3. When most students have completed the first two problems, have the class discuss what strategies they are using. If no students move to the symbolic form, lead the discussion in that direction.(Refer to *Background Information).*
4. Allow students to solve the remaining problems before having the class share their solutions and strategies.

Discussion

1. What are some strategies you used to solve this problem? [guess and check, working backwards]
2. How can you translate what you did into algebraic symbols?

* Reprinted with permission from *Principles and Standards for School Mathematics,* 2000 by the National Council of Teachers of Mathematics. All rights reserved.

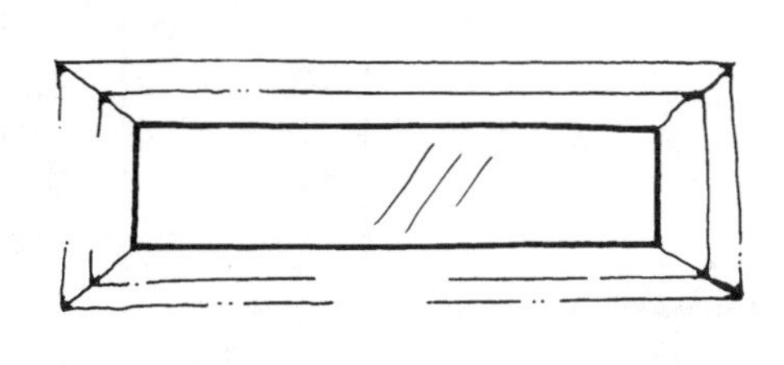

The dimensions for an order of windows has been lost at Panes of Glass Windows. As an employee, you are given the invoice. Determine the dimensions of the window from the price, the window type, and your knowledge that the price of a window is set at $2.00 a square foot.

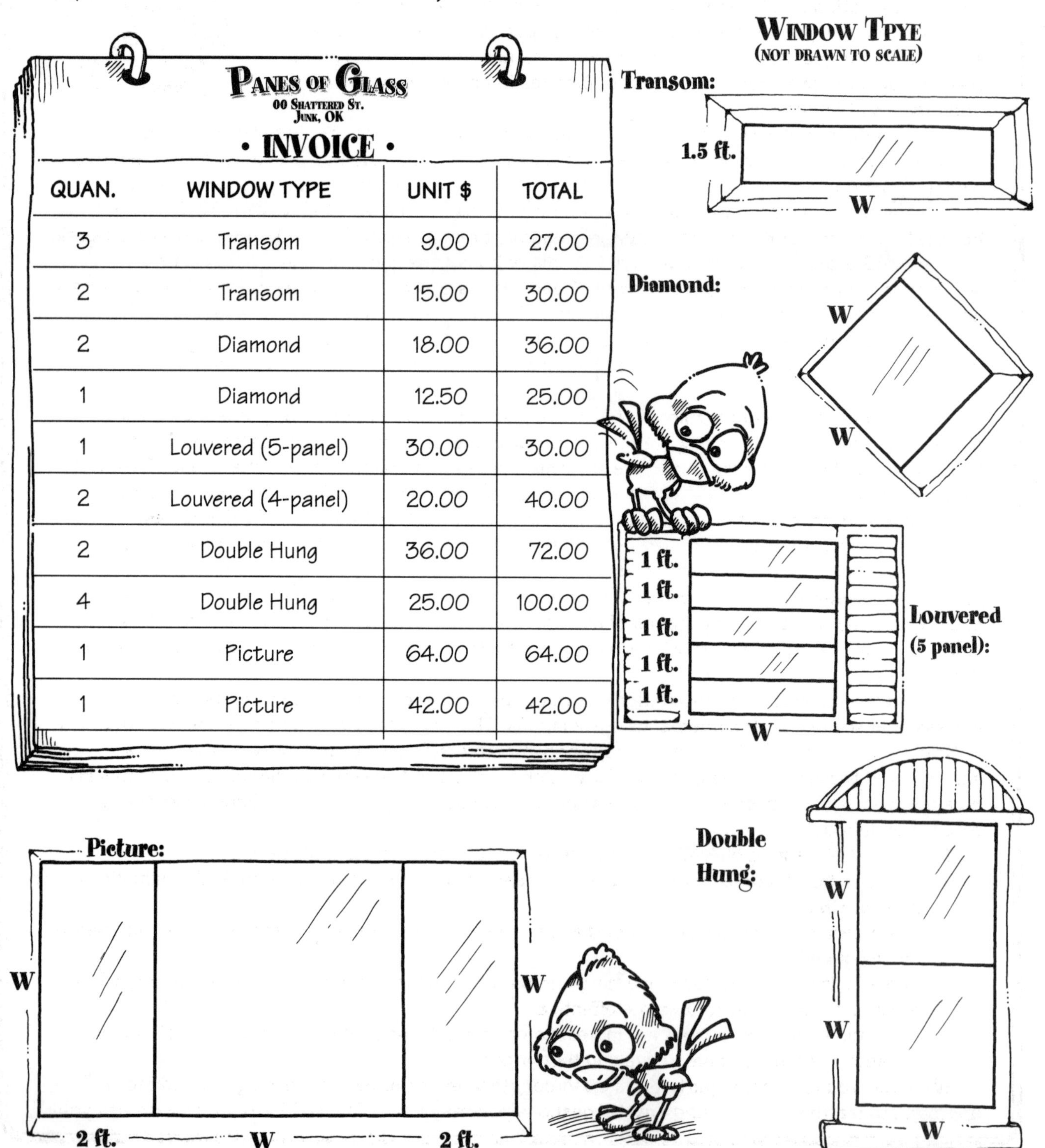

PANES OF GLASS
00 Shattered St.
Junk, OK

• INVOICE •

QUAN.	WINDOW TYPE	UNIT $	TOTAL
3	Transom	9.00	27.00
2	Transom	15.00	30.00
2	Diamond	18.00	36.00
1	Diamond	12.50	25.00
1	Louvered (5-panel)	30.00	30.00
2	Louvered (4-panel)	20.00	40.00
2	Double Hung	36.00	72.00
4	Double Hung	25.00	100.00
1	Picture	64.00	64.00
1	Picture	42.00	42.00

The AIMS Program

AIMS is the acronym for "**A**ctivities **I**ntegrating **M**athematics and **S**cience." Such integration enriches learning and makes it meaningful and holistic. AIMS began as a project of Fresno Pacific University to integrate the study of mathematics and science in grades K-9, but has since expanded to include language arts, social studies, and other disciplines.

AIMS is a continuing program of the non-profit AIMS Education Foundation. It had its inception in a National Science Foundation funded program whose purpose was to explore the effectiveness of integrating mathematics and science. The project directors in cooperation with 80 elementary classroom teachers devoted two years to a thorough field-testing of the results and implications of integration.

The approach met with such positive results that the decision was made to launch a program to create instructional materials incorporating this concept. Despite the fact that thoughtful educators have long recommended an integrative approach, very little appropriate material was available in 1981 when the project began. A series of writing projects ensued, and today the AIMS Education Foundation is committed to continuing the creation of new integrated activities on a permanent basis.

The AIMS program is funded through the sale of books, products, and staff development workshops, and through proceeds from the Foundation's endowment. All net income from programs and products flows into a trust fund administered by the AIMS Education Foundation. Use of these funds is restricted to support of research, development, and publication of new materials. Writers donate all their rights to the Foundation to support its on-going program. No royalties are paid to the writers.

The rationale for integration lies in the fact that science, mathematics, language arts, social studies, etc., are integrally interwoven in the real world, from which it follows that they should be similarly treated in the classroom where students are being prepared to live in that world. Teachers who use the AIMS program give enthusiastic endorsement to the effectiveness of this approach.

Science encompasses the art of questioning, investigating, hypothesizing, discovering, and communicating. Mathematics is a language that provides clarity, objectivity, and understanding. The language arts provide us with powerful tools of communication. Many of the major contemporary societal issues stem from advancements in science and must be studied in the context of the social sciences. Therefore, it is timely that all of us take seriously a more holistic method of educating our students. This goal motivates all who are associated with the AIMS Program. We invite you to join us in this effort.

Meaningful integration of knowledge is a major recommendation coming from the nation's professional science and mathematics associations. The American Association for the Advancement of Science in *Science for All Americans* strongly recommends the integration of mathematics, science, and technology. The National Council of Teachers of Mathematics places strong emphasis on applications of mathematics found in science investigations. AIMS is fully aligned with these recommendations.

Extensive field testing of AIMS investigations confirms these beneficial results:

1. Mathematics becomes more meaningful, hence more useful, when it is applied to situations that interest students.
2. The extent to which science is studied and understood is increased when mathematics and science are integrated.
3. There is improved quality of learning and retention, supporting the thesis that learning which is meaningful and relevant is more effective.
4. Motivation and involvement are increased dramatically as students investigate real-world situations and participate actively in the process.

We invite you to become part of this classroom teacher movement by using an integrated approach to learning and sharing any suggestions you may have. The AIMS Program welcomes you!

AIMS Education Foundation Programs

Practical proven strategies to improve student achievement

When you host an AIMS workshop for elementary and middle school educators, you will know your teachers are receiving effective usable training they can apply in their classrooms immediately.

Designed for teachers—AIMS Workshops:

- Correlate to your state standards;
- Address key topic areas, including math content, science content, problem solving, and process skills;
- Teach you how to use AIMS' effective hands-on approach;
- Provide practice of activity-based teaching;
- Address classroom management issues, higher-order thinking skills, and materials;
- Give you AIMS resources; and
- *Offer college (graduate-level) credits for many courses.*

Aligned to district and administrator needs—AIMS workshops offer:

- Flexible scheduling and grade span options;
- Custom (one-, two-, or three-day) workshops to meet specific schedule, topic and grade-span needs;
- Pre-packaged one-day workshops on most major topics—only $3,900 for up to 30 participants (includes all materials and expenses);
- Prepackaged *week-long* workshops (four- or five-day formats) for in-depth math and science training—only $12,300 for up to 30 participants (includes all materials and expenses);
- Sustained staff development, by scheduling workshops throughout the school year and including follow-up and assessment;
- Eligibility for funding under the Eisenhower Act and No Child Left Behind; and
- Affordable professional development—save when you schedule consecutive-day workshops.

University Credit - Correspondence Courses

AIMS offers correspondence courses through a partnership with Fresno Pacific University.

- *Convenient* distance-learning courses—you study at your own pace and schedule. No computer or Internet access required!

The tuition for each three-semester unit graduate-level course is $264 plus a materials fee.

The AIMS Instructional Leadership Program

This is an AIMS staff-development program seeking to prepare facilitators for leadership roles in science/math education in their home districts or regions. Upon successful completion of the program, trained facilitators become members of the AIMS Instructional Leadership Network, qualified to conduct AIMS workshops, teach AIMS in-service courses for college credit, and serve as AIMS consultants. Intensive training is provided in mathematics, science, process and thinking skills, workshop management, and other relevant topics.

Introducing AIMS Science Core Curriculum

Developed in alignment with your state standards, AIMS' Science Core Curriculum gives students the opportunity to build content knowledge, thinking skills, and fundamental science processes.

- *Each* grade specific module has been developed to extend the AIMS approach to full-year science programs.
- *Each* standards-based module includes math, reading, hands-on investigations, and assessments.

Like all AIMS resources these core modules are able to serve students at all stages of readiness, making these a great value across the grades served in your school.

For current information regarding the programs described above, please complete the following:

Information Request

Please send current information on the items checked:

____ *Basic Information Packet* on AIMS materials ____ Hosting information for AIMS workshops
____ *AIMS Instructional Leadership Program* ____ AIMS Science Core Curriculum

Name ______________________ Phone ______________________

Address ____
Street City State Zip

AIMS Program Publications

Actions with Fractions 4-9
Awesome Addition and Super Subtraction 2-3
Bats Incredible! 2-4
Brick Layers 4-9
Brick Layers II 4-9
Chemistry Matters 4-7
Counting on Coins K-2
Cycles of Knowing and Growing 1-3
Crazy about Cotton Book 3-7
Critters K-6
Down to Earth 5-9
Electrical Connections 4-9
Exploring Environments Book K-6
Fabulous Fractions 3-6
Fall into Math and Science K-1
Field Detectives 3-6
Finding Your Bearings 4-9
Floaters and Sinkers 5-9
From Head to Toe 5-9
Fun with Foods 5-9
Glide into Winter with Math & Science K-1
Gravity Rules! Activity Book 5-12
Hardhatting in a Geo-World 3-5
It's About Time K-2
It Must Be A Bird Pre-K-2
Jaw Breakers and Heart Thumpers 3-5
Problem Solving: Just for the Fun of It! 4-9
Looking at Geometry 6-9
Looking at Lines 6-9
Machine Shop 5-9
Magnificent Microworld Adventures 5-9
Marvelous Multiplication and Dazzling Division 4-5
Math + Science, A Solution 5-9
Mostly Magnets 2-8
Movie Math Mania 6-9
Multiplication the Algebra Way 4-8
Off The Wall Science 3-9
Our Wonderful World 5-9
Out of This World 4-8
Overhead and Underfoot 3-5
Paper Square Geometry:
The Mathematics of Origami
Puzzle Play: 4-8
Pieces and Patterns 5-9
Popping With Power 3-5
Primarily Bears K-6
Primarily Earth K-3
Primarily Physics K-3
Primarily Plants K-3
Proportional Reasoning 6-9
Ray's Reflections 4-8
Sense-Able Science K-1
Soap Films and Bubbles 4-9
Spatial Visualization 4-9
Spills and Ripples 5-12
Spring into Math and Science K-1
The Amazing Circle 4-9
The Budding Botanist 3-6
The Sky's the Limit 5-9
Through the Eyes of the Explorers 5-9
Under Construction K-2
Water Precious Water 2-6
Weather Sense:
Temperature, Air Pressure, and Wind 4-5
Weather Sense: Moisture 4-5
Winter Wonders K-2

Spanish/English Editions*

Brinca de alegria hacia la Primavera con las Matemáticas y Ciencias K-1
Cáete de gusto hacia el Otoño con las Matemáticas y Ciencias K-1
Conexiones Eléctricas 4-9
El Botanista Principiante 3-6
Los Cinco Sentidos K-1
Ositos Nada Más K-6
Patine al Invierno con Matemáticas y Ciencias K-1
Piezas y Diseños 5-9
Primariamente Física K-3
Primariamente Plantas K-3
Principalmente Imanes 2-8

* All Spanish/English Editions include student pages in Spanish and teacher and student pages in English.

Spanish Edition

Constructores II: Ingeniería Creativa Con Construcciones LEGO® (4-9)
The entire book is written in Spanish. English pages not included.

Other Science and Math Publications

Historical Connections in Mathematics, Vol. I 5-9
Historical Connections in Mathematics, Vol. II 5-9
Historical Connections in Mathematics, Vol. III 5-9
Mathematicians are People, Too
Mathematicians are People, Too, Vol. II
Teaching Science with Everyday Things
What's Next, Volume 1, 4-12
What's Next, Volume 2, 4-12
What's Next, Volume 3, 4-12

For further information write to:
AIMS Education Foundation • P.O. Box 8120 • Fresno, California 93747-8120
www.aimsedu.org/ • Fax 559•255•6396